A TALE OF TIME

The Journey of Space and Time

Understanding the very beginning of our universe and time through a beautiful journey

ADITYA SINGHA

Copyright notice and disclaimer

All rights of this book including the cover art belong to the author. Any means of reproduction of the material is prohibited without the written consent of the author. The copyright protection is as per the Berne convention and local laws of the concern country.

The book is a non-fiction work of science and all materials mentioned in the book are based on scientific data and actual physical observation based on logical and analytical reasoning, in no way the author wants to show any disrespect to any person's belief.

Science gave mankind the wings to fly and the power to survive against the forces of nature, but it was nature who gave us time to evolve in the first place. Science is nature's way to unveil itself in front of us.

Science is the essence of the universe, its study is the process to understand it and getting closer to the true universal entity.

Aditya Singha

Content

Topic	Page Number
Introduction	8
Time- the guardian of universe	13
Understanding universe	29
The domain of space and time	44
The birth of legends-1	62
The birth of legends -2	88
A humble journey	101
Time – a blessing to mankind	114
A symphony of beauty	131
Dark Matter – a mysterious anomaly	142
Time travel- a journey to remember	149
Parallel universe- a closer neighbour	162
Our future- living the dream	177

Preface

The most mysterious questions are the ones which led us to dive deep into our emotions and look for the unknown, to search for our existence and to know the answer to the question of our creation. As humans, since the dawn of civilization some of the most prominent questions that drive mankind to strive for more knowledge were the questions of our existence and the origin of our universe. Various works are done of different contexts including religions, philosophy and science. Time played a significant role in all these events which at the end led to the development of mankind and the civilisation as we know today. In this book we will understand the true meaning of time in a scientific way and its role in creating life as we know. We will take a journey to unravel the significance in the creation of our beautiful planet and how life

emerged through an evolutionary journey of billions of years. We will understand the various timelines which favours our existence, possibility of time travel and will understand the mysteries of the presence of life on other planets and why we have not yet discovered any.

The book is designed in a simple and elegant way to explain some of the most fascinating scientific events as simply as it can be, so that everyone of all ages can enjoy and understand the beauty and importance of time in our creation and overall existence. By knowing some basic scientific mysteries about our origin will definitely put a new outlook towards our individual life and towards a more sustainable growth which is a high demand of our time.

I would like to state a disclaimer that this book in no way wants to hurt anybody's feelings or religious beliefs. Every statement whatsoever is made based on actual data, reasoning and scientific hypothesis.

So let us begin this journey of billions of years of evolution, one step at a time.

ADITYA SINGHA

Date-18th April 2021
India   Author

Introduction

We are a part of a great and beautiful creation of nature which gave us our planet earth, our sun including billions of other stars, galaxies and countless other celestial bodies. All of them are bound to each other with natural gravitational forces. But nature doesn't stop fascinating us there; it includes its amazing gifts in many forms including creation of life. Through ages of exploration done by our ancestors we have bestowed ourselves with the power of knowledge which beacon our path beyond superstition and have helped us to reach the modern age of science and technology. On this topic, one of the greatest scientist of all time, Sir Isaac Newton had made a quote *"If I have seen further than others, it is by standing upon the shoulders of giants"*.

Without knowledge humans are like a ship without a sail, without any destination, left to sink in the vast ocean of myths. Science has always taught us to grow and understand things around us. This is how we understand the motion of planets, Euclidean geometry, and the formation of life and the concept of evolution to name a few.

This book is written to understand us regarding how our universe was created and how all its constituent bodies which include stars, planets, and nebulas are created. Along with that the most important part that this book covers is the concept of time and its flow in the space time continuum. Many great scientists around the world have made several claims regarding time travel which may confuse a common person so this book is developed to make sure we understand what time really is and how to travel through time. As we move through this book we will learn some really amazing

concepts in a much simpler way so that every person of all ages can learn science and enjoy it to its full potential.

In today's time it is shameful that although we live in a society where science enlightens us, technology empowers us but still we are fighting under the shadows of blunt talks of superstition, racism and hatred towards different communities. But whenever we look at our nature we can find a comforting idea that nature never differentiate between us, the sun that lightens us, the air we breathe, the water we drink, the colour of our blood is not differentiated by which religion we follow, which country we are born in. In a way universe join us together through the spirit of nature, a beautiful quote by Ralph Waldo Emerson describe beautifully how lucky we are to be graced by this bond of nature. The quote is as follows, "*If the stars should appear one night in a thousand years, how would men believe and adore; and preserve for many generations the remembrance of the city of God which had been shown! But every night come out these envoys of beauty, and light the universe with their admonishing smile"*

With the true knowledge of our creation we get the simplicity of life and become humble in our approach towards others. In a world of chaos where different types of evil have taken over and divided humanity, maybe understanding the cradle of our existence would remind us what it is meant to be human; to be close to the universe and to rekindle the meaning of humanity.

Understanding how insignificant yet meaningful our creation is and how delicate the balance of life is, we can achieve a new sense of perspective to save and preserve our only home, our planet. I do hope by the end of this book the readers will have obtain not only meaningful scientific insight but will gain the stories of incredible coincidences that al led us to the present day, no matter what we call it,

the mysteries of universe is always fascinating and a mirror to our existence. We will begin our journey, by understanding the creation of time itself and thereby the birth of our universe. We will then take billions of years of steps in few chapters to understand the beautiful celestial bodies, their creation and what significant powers of universe they hold and then to the most fascinating question which is depicted in every sci-fi movies and fantasy tales, the concept of time travel and is it possible. Without further deferment, let's start this humbling voyage.

Chapter 1

Time - the guardian of universe

The word which we use several times in a day is the word which is most delusional and its real meaning is very confusing to the common people. If you ask any person what time the most probable answer that you are going to receive is that it's the duration of an event and they are not wrong, as that is what we are told from our childhood but the question is whose duration and is this most applicable definition of time?

If we look the definition of the international standard unit of time which is second we will find it as follows :- *"**the duration of 9 192 631 770 periods of the radiation corresponding to the transition between the two hyperfine levels of the ground state of the caesium 133 atom"*. Now again for a person who is not related to physical science this definition will not make any sense, so let bring it down to layman's term. Basically it says that if we take a caesium (a radioactive element of the periodic table) atom with a particular mass number and allow its radiation level to transit between two lower energy state of caesium for 9,192,631,770 times (which is measured by a series of sophisticated experiments) we will come to a moment of duration which the SI system define as one second. If we move back in time when science was not as developed as it is today the definition of second was defined as the fraction 1/86,400 of the mean solar day but the problem with this definition is that it did not define what a solar day is. If we define solar day based on earth rotation it will be contradictory as earth rotation have certain irregularities based on our calculation, we will discuss this in a minute and moreover other planets in our universe does not follow earth's rotation pattern so this definition cannot be accepted as an international standard.

Now before we precede any further I have to describe to you why there is an irregular pattern in earth's rotation which is one of the reasons why the earlier definition of second was supposed to be changed. Well the reason is the change of speed and orientation of earth's motion around the sun but let's elaborate the situation.

1. We all have studied in our school life that earth rotates around the sun in elliptical orbits which is true but if you understand Euclidean geometry you will understand the direction of earth's rotation is accompanied by a tilt in its axis of rotation. As of May of 2017 this tilt is 23.5^0 which will change in a cycle that averages around 40,000 years within a range of 22.1 and 24.5 degrees. This change in tilting position has an effect on the rotation period of earth which although is very subtle can affect precise calculation for the definition of second.

2. Whenever any person asks us what is the shape of earth we always give a very straight forward answer that it is spherical but that is not exactly correct geometrically speaking, let's see why. Well in geometry a sphere is one which in which all points on its locus of surface is in equal distance from its centre. Whereas this is not true in the case of earth, near the equator the diameter is (12,756.27 km (7,926.38 mi)) and for the pole the same value is (12,713.56 km (7,899.84 mi)) with a difference of about 42.77 Km. this difference is known as the equatorial bulge. This is due to the rotation of earth and for which it is not exactly spherical. The shape is sometime refers to us oblate spheroid or in a more theoretical sense known as Geoid shape, considering only the effect the gravitational force of the earth and its rotation while ignoring all other forces of nature.

3. We know that earth is orbiting around its elliptical orbit because of the gravitational pull from the sun. The moon also has a pull but since its mass is too less in comparison to earth that the only effect it can create is tidal waves. However the pull from sun is quite effective but as we know from Newton's gravitational law, that the force of gravitation is indirectly proportional to the distance between the two bodies and as a result as the planet moves away from the sun (the winter season) the force decreases and vice versa when it comes closer to it. Now practically speaking this change is not quite effective in comparison to the huge gravitational force of attraction that the sun possess but still while defining minute definition like that of time itself it lies a big factor.

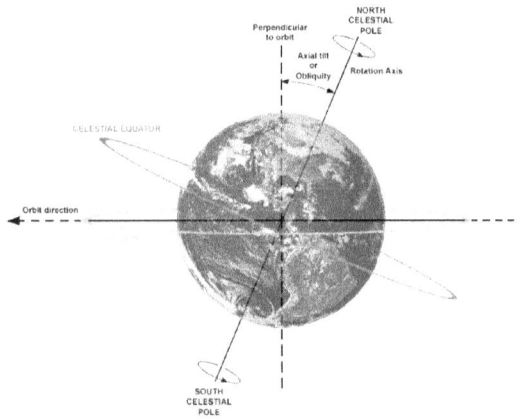

Fig. - 1A:

Earth's **axial tilt** is about 23.4°. It oscillates between 22.1° and 24.5° on a 41,000-year cycle and is currently decreasing.

Credit - Wikipedia creative commons license distribution.

FIG. 1B

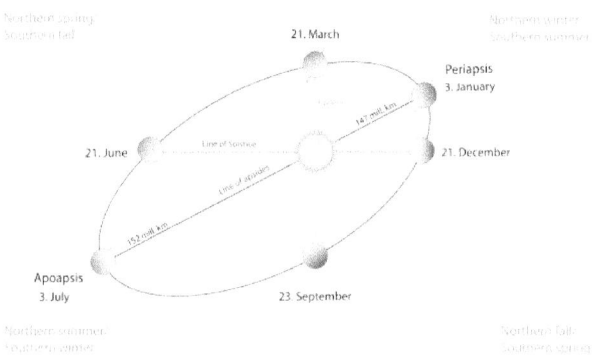

Elliptical Motion of earth around sun
Credit- Wikipedia creative commons license

These are some of the reason why the international scientific community decided to alter the definition of the standard of time, viz.: second in a way which can be more easily definable and applicable by the universals laws of physics without any boundaries. This is how we get our current definition of second or more specifically an understanding of time itself.

So till now we have learned that time as intercepted by normal human brain can be said to be duration of an event in the physical world abiding the laws of nature. However one of the greatest physicist, Albert Einstein in his paper of special and general theory of relativity defined time in an entirely different way. This we will learn in an upcoming chapter. He viewed time like a stream of water which is unidirectional in its path of flow and this was an important conceptual reason why many believed in time travel in the first place. The idea was if time moves as a river flows

downstream, then isn't it possible to move upstream in a river, so why not in time, more specifically time travel.

But is time travel and its concept so simple and if yes why haven't we achieved it yet and if it is possible what are its scope?
 There are many such questions which might be in your mind right now but we are just in the beginning of a very interesting journey ahead where we will learn few basic concepts of our space, our planet and even our very existence as a life form. Understanding these concepts will help us to understand time more closely and not just as a measure of duration of an event.
For time being (no pun intended :)), let us keep in mind the general definition of time as mention in the SI system and let's explore more.
One of the best parameter to measure time and distance is done using speed of light but have you ever thought why we need a

universal standard to measure time, well this was first thought by none other than Newton. But let's understand this with an example, let's consider you are inside a train moving at 120kmph (74.56 MPH) and you are playing with a ping pong ball. So you threw the ball to the ceiling and catch it once it fall, for you in that one second the ball makes no lateral displacement so net velocity is zero. Now let's consider your friend who is standing on the platform, from his point of view that same ping pong ball has moved laterally almost 33 meters because of the speed of the train. This is what Newton worried because if we want to find the mysteries of our universe than we have to at least get its origin timeline correct which is only possible, in fact before Newton even Aristotle also believed in the idea of an absolute time which will not change with change in reference frame. So the only reasonable idea was to find a universal reference which will not suffer such a change based on change in observer's reference frame and that is the speed of light

which remains constant through any frame, the reason will be discussed in detail in chapter 3.

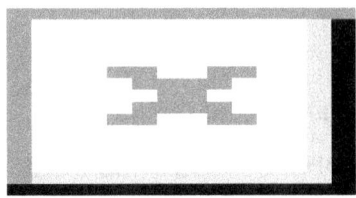

Fig. – 1C: The motion of the ping pong ball

In 1887 Albert Michelson (Nobel Laureate in physics) and Edward Morley conducted a very careful experiment at the Case school of Applied Science in Cleveland. They compared the speed of light in the direction of the earth's motion with that at right angles to the earth's motion and it was one of the revealing experiment to find that the speed match. Understand the reason why this experiment was so important and why later in 1905 when Einstein published his paper on theory of special relativity the result was so groundbreaking, it's because no we have a set of physical laws which will hold well throughout the universe, well except black holes. This was the first step in the long exploration of our final frontier of space and to unravel its mysteries.

The remarkable idea lies when we compare Newtonian physics with that of relativity, let's have a look. As per Newton's, two different observers when seeing a beam of light would agree on the duration of the travel since speed of light will remain constant but may not agree on the distance it travelled because space is not absolute but this mean for different observers the speed of light would be different but with theory of relativity this idea changed and that give us the idea of there is no concept of absolute time, for each observer it will be different and this was in fact the first step which took mankind closer to understand the concept of time travel.

But one of the most fascinating aspects of time is that, it is actually the fourth dimension of our universe and in reality also we have quite an understanding of what it is in our daily lives. Let us understand this with an example, so let's say you have asked your friend to meet you at your office to pick you up, in order to actually find you your friend will need GPS location which will rely on the three dimensional co-ordinates namely the length ,width and height. Actually using these coordinates only we can place anyone anywhere on this universe, from a small insect to a giant star. But there is vital information which is missing, your friend needs the details of the time when to pick you up, else it's going to be quite difficult.

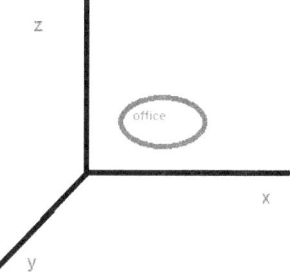

Fig. 1D: *the concept of three dimensions is incomplete without the knowledge of the fourth dimension.*

Although we really don't think about it but even subconsciously we are all using time as our fourth dimension. This information is vital not only in our daily lives but also when understanding universe, because it provides information regarding the timeline and expand our knowledge base to decipher the mysteries of universe.

All those beautiful stars we all see in our night sky is the testimony of our past, when we say any star is 5 million light years old, all we want to say is we are seeing an image of the star which is 5 million year old. So, can we use this concept in time travel? To understand that we must understand the concept of space and as we move through different chapters of the book we will have a better understanding of this concept.

Time acts as a book keeper of the past events of our universe, it's the guardian angel of our present because it will keep these memories of our lives for a new life form somewhere few million later so that when they see the earth they observe our present to shape their future. But time is incomplete without space, for they are compliment to each other. So far this has been just a small introduction to time, let's understand the canvas of our story, our universe which will give us a better perspective of the importance of time and we will learn the tales of our past, our creation which eventually created our present. It's said that past is a story and future is a mystery but our present is a gift which is why we call it present. So we should admire the events which led us to our present, so let's dive into the story of our universe.

Chapter 2

Understanding Universe

Astronomy is a humbling subject which teaches us the morale of humanity and in today's world of chaos and people with lesser knowledge showcasing their pride to feed their ego, none other than astronomy can take us to the true meaning of life and its purpose. But to understand the vastness of this topic we have to begin our journey by knowing the essence of our existence in this canvas of space and time.

We got our first self-aware how small we are in the cosmos when the Voyager-1 spacecraft took the first picture of earth from Saturn, also known as the "Pale Blue Dot". We are not special in anyway, yet we are created from the cosmos, to quote Carl Sagan, "We're made of star stuff. We are a way for the cosmos to know itself". This chapter is both scientifically approachable as well as it deals with the philosophical idea of our existence.

Knowledge is an ever-lasting phenomenon and no matter how much you think you have learned there is always more to it, hence the

first point is to learn humility. With that begin said, have a look at this image.

Fig. 2A: A historical pillar engraving from the Mauryan period, India. The picture depicts blind people's interpretation of an elephant

I took this picture few years back at the national history museum, Kolkata. The pillar is from medieval India and it denotes a very important concept which is applicable not only to cosmology and space exploration but also to life. We interpret things based on our past experiences or knowledge and that cloud our mind with prejudicial ideas which is never a good thing. The blind men who touched the tail of the elephant will have a different image of the animal then those who touched his leg or ear.

This concept is very important in space exploration as well. When we are defining space and time and expanding our ideas to what life is, we must not cloud our ideas to that of how life is on our planet, based on this the first widely accepted proposal was given by Dr. Frank Drake in 1961, popularly known as Drake equation. We won't go too deep into its mathematical formula but the equation consider several factors while determining if life is possible on other planets. This includes a critical factor of time among many other conditions. The equation is as follow:

$$N = N_s \times F_p \times F_1 \times F_i \times L_c / L_s$$

N is the number of civilizations in the Milky Way today.

N_s is the number of stars in the Milky Way.

F_p is the fraction of stars with habitable planets.

F_1 is the fraction of habitable planets with life.

F_i is the fraction of life-bearing planets where intelligent civilizations arise.

L_c is the typical life time of a civilization in years.

L_s is the typical life time of a star (10 billion years for Sun-like stars).

But before we move to the concept of life, we have to understand the beginning of this canvas or was there any beginning at all? Was it a cyclic event or are we the first to witness a series of event? The answer to all these questions lies in the epicentre of this chapter; let's have a look in the origin of our universe.

Throughout history there have been two schools of thought on the creation of our universe. Some people were against the idea of creation of universe because according to them it always existed, but these philosophical people could never provide a rationale reason for this, such people include great Greek philosopher, Aristotle and German philosopher, Immanuel Kant. They did this because then they don't have to consider any divine interpretation in the idea of creation of universe.

Then the second school of thought which is that of creation is bifurcated into two parts, one who consider it based on some creator while the other is based on facts and research, the scientific one.

In 1915 after the publication of Einstein's general theory of relativity we got a more unified approach of space and time. And since neither of them are absolute hence they cannot exist before the creation of universe, hence it's for sure that our universe had a creation. The more scientific approach to this concept can be seen in the next chapter.

Postulate 1: THE BIG BANG THEORY

The most popular theory regarding the creation of universe is the big bang theory which gives a well-defined unified version of its creation. But there was a major debate over this topic which is the existence of time. As per the big bang time did not existed before this event and this is little hard to explain when this theory was first proposed in 1920s by a Belgian priest named Georges Lemaitre , when he theorized that the universe began from a single primordial atom. The idea received major boosts from Edwin Hubble's observations that galaxies are speeding away from us in all directions, as well as from the 1960s discovery of cosmic microwave radiation—interpreted as echoes of the big bang—by Arno Penzias and Robert Wilson.

Further work has helped clarify the big bang's tempo. Here's the theory: In the first 10^{-43} seconds of its existence, the universe was very compact, less than a million billion billionth the size of a single atom. It's thought that at such an incomprehensibly dense, energetic state, the four fundamental forces—gravity, electromagnetism, and the strong and weak nuclear forces—were forged into a single force, but our current theories haven't yet figured out how a single, unified force would work. To pull this off, we'd need to know

how gravity works on the subatomic scale, but we currently don't. This still did not help us understand when and how time suddenly came into existence. But then during the golden era for theoretical physicists in 1990s, the advent of string theory along with its concept of multiple dimensions helped us to understand the creation of universe. These postulates were further confirmed with the help of the Large Hadron Collider in CERN laboratory, Switzerland. Although string theory is a vast topic still as we progress into this book, we will decipher a basic of this mystery of string theory.

But the big bang was the first step, after the explosion as time passed and matter cooled, more diverse kinds of particles began to form, and they eventually condensed into the stars and galaxies of our present universe. By the time the universe was a billionth of a second old, the universe had cooled down enough for the four fundamental forces to separate from one another. The universe's fundamental particles also formed. It was still so hot, though; that these particles hadn't yet assembled into many of the subatomic particles we have today, such as the proton. As the universe kept expanding, this piping-hot primordial soup—called the quark-gluon plasma—continued to cool. Some particle colliders, such as CERN's Large Hadron

Collider, are powerful enough to re-create the quark-gluon plasma.

Radiation in the early universe was so intense that colliding photons could form pairs of particles made of matter and antimatter, which is like regular matter in every way except with the opposite electrical charge. It's thought that the early universe contained equal amounts of matter and antimatter. But as the universe cooled, photons no longer packed enough punch to make matter-antimatter pairs. So like an extreme game of musical chairs, many particles of matter and antimatter paired off and annihilated one another. Somehow, some excess matter survived—and it's now the stuff that people, planets, and galaxies are made of. Our existence is a clear sign that the laws of nature treat matter and antimatter slightly differently. Researchers have experimentally observed this rule imbalance, called CP violation, in action. Physicists are still trying to figure out exactly how matter won out in the early universe.

Within the first second, it was cool enough for the remaining matter to coalesce into protons and neutrons, the familiar particles that make up atoms' nuclei. And after the first three minutes, the protons and neutrons had assembled into hydrogen and helium nuclei. By mass, hydrogen was 75 percent of the early universe's matter, and helium was 25

percent. The abundance of helium is a key prediction of big bang theory, and it's been confirmed by scientific observations. Despite having atomic nuclei, the young universe was still too hot for electrons to settle in around them to form stable atoms. The universe's matter remained an electrically charged fog that was so dense; light had a hard time bouncing its way through. It would take another 380,000 years or so for the universe to cool down enough for neutral atoms to form—a pivotal moment called recombination. The cooler universe made it transparent for the first time, which let the photons rattling around within it finally zip through unimpeded. We still see this primordial afterglow today as cosmic microwave background radiation, which is found throughout the universe. The radiation is similar to that used to transmit TV signals via antennae. But it is the oldest radiation known and may hold many secrets about the universe's earliest moments. But there were various evolution stages which we will discuss in a later chapter.

Postulate 2: Arrow of time

In order to understand this postulate we first have to understand the idea of entropy. In thermodynamics we can simply state entropy as the measure of randomness in the system. This mean as the randomness increases, entropy also increase along with it.

Energy, Entropy, the 2ⁿᵈ law of Thermodynamics

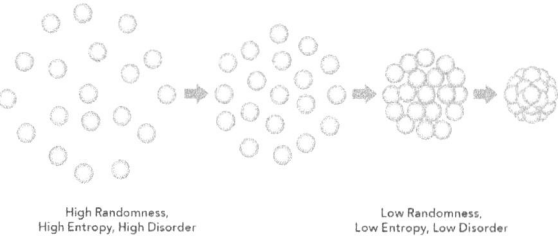

High Randomness, High Entropy, High Disorder   Low Randomness, Low Entropy, Low Disorder

Fig 2B: The flow of entropy in a thermodynamic process. Here a decrease is depicted.

Now this theory works on the idea that time can only flow in one direction, which is in the future. With the passage of time randomness also increases and hence there is an increase in entropy. But let us understand with a simple scenario. Let's say you went for shopping, and by mistake you drop some money. If you immediately start looking for it, chances are you will find it soon but with each passage of moment it become increasingly difficult. The same is true with universe, as it expands the randomness increase.

Entropy represents this randomness of a system which is shown as the unavailability of energy in a system to do work. By randomness I mean how unorganized a system is, for example let's say there are 100 students in a classroom which in this case is our system. Now all the students are wearing the same uniform and have no name tags, the only way possible to identify them is by looking at their face. Now if they are all calm & seating in their respective seats then one can easily identify any one of them without much of a trouble, but if they are all running in the classroom, randomly, playing or creating havoc then it is very hard comparatively to identify one of them that easily. Now just replace the students by molecules or any other particle and the classroom by any other related system, we have our definitive idea of entropy. In fact, our universe behaves as a giant system on its own.

But as a reader you might be thinking that in the definition it is mentioned that it is the amount of energy which is unavailable for work, how can randomness in a system relate to that. Well my friend let's go back to our example of those students to understand it. Initially we have said that those students are quietly sitting in their respective seats, which means randomness is zero (for molecules or electrons this is never true because they are always in a constant state of motion so there

will be some randomness). Now in that situation if the teacher asks the students to answer a question almost all of them can do that because all of them are paying attention and hence are available for the work given by the teacher (Lets consider them to be very discipline students). Now later we have mention that those children are playing inside the classroom creating havoc and a lot of randomness is there in the system, now if the teacher ask to give the answer to that same question only a few students will be able to answer it since most other kids are busy playing. This shows how the randomness in the system can lead to a decrease in participant for a work. This is where the idea of real life system comes, if there is too much randomness in a system then the amount of energy available to do work will also decrease. Remember that our universe doesn't exactly participate like a high school; this example was just given to make the scenario much more relatable. Just for now keep this idea in mind that our universe act as a giant system on its own.

Now let's move to enthalpy. It is pretty much defined as the total amount of energy available

for the work. Thermodynamically speaking it is equal to the total heat content of the system given by the total internal energy and the product of pressure and volume of the content of the system.

Mathematically,

$$H = U + PV,$$

Where H is enthalpy, U is internal energy and P, V represents the pressure and volume.

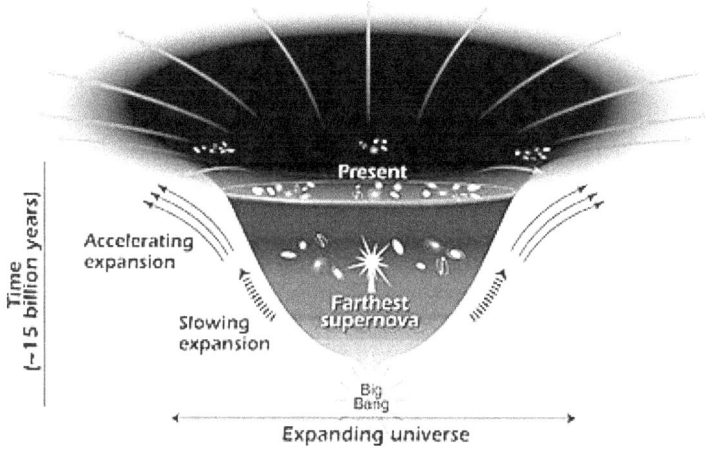

This diagram reveals changes in the rate of expansion since the universe's birth 15 billion years ago. The more shallow the curve, the faster the rate of expansion. The curve changes noticeably also at 7.5 billion years ago, when objects in the universe began flying apart at a faster rate. Astronomers theorize that the faster expansion rate is due to a mysterious, dark force that is pushing galaxies apart.

Fig 2C: Expansion of universe through the course of time

In the year 2005, cosmologist of university of Chicago Sean Carroll and his grad student Jennifer Chen published a paper on this topic. The research indicates that our universe has been expanding since the last 13 billion years and will continue to do so. Even if big bang started it, what were the criteria that set up the big bang?

Carroll found that process hidden inside one of the strangest and most exciting recent elaborations of the Big Bang theory. In 1984, MIT physicist Alan Guth suggested that the very young universe had gone through a brief period of runaway expansion, which he called "inflation," and that this expansion had blown up one small corner of an earlier universe into everything we see. In the late 1980s Guth and other physicists, most notably Andrei Linde, now at Stanford, saw that inflation might happen over and over in a process of "eternal inflation." As a result, pocket universes much like our own might be popping out of the uninflected background all the time. This multitude of universes was called, inevitably, the multiverse. This looks like a science fiction but in reality it's very much theoretical, in fact one of the key ideas if the universe keep on expanding is the ultimate death of universe by formation of several super massive black holes. They once again will collapse to form another big bang. So does that mean we had another universe before ours was created or are there several other multiverse still in existence as in parallel universe? Although all answers are still not in our hand, we do have sufficient ideas to plot the events that led the creation of our universe including the concept of dark matter.

Chapter 3

The Domain of space and time

In 1915 after Einstein's theory of relativity was unveiled to the world, it changed people perception of time and space. We can't suddenly jump to this relationship without actually understanding the fundamentals behind them. So let's start from the basics, humans have evolved on planet earth only a few thousand years ago and in comparison to the age of the universe which is billions of years, our time on this planet is quite insignificant. However, the postulates and theories are based on rigorous mathematical calculations and it gave us a brief idea on what time is and how is it related to space. So, let's have a look at it with the help of the following example.

Imagine you and your friend is holding a handkerchief in such a way that each one of you has two ends of the handkerchief. So, the handkerchief is now flat and if you place a small tennis ball in the middle of it, then this will represent any celestial object in the fabric of space and time. Most probably it will look something like this if you have held the handkerchief properly. Now the fascinating thing about this demonstration shows how gravity directly interferes with space and time. The heavier the object is the more it will bend the handkerchief or in other words the heavier the mass of any celestial object will be the more it can bend space and time.

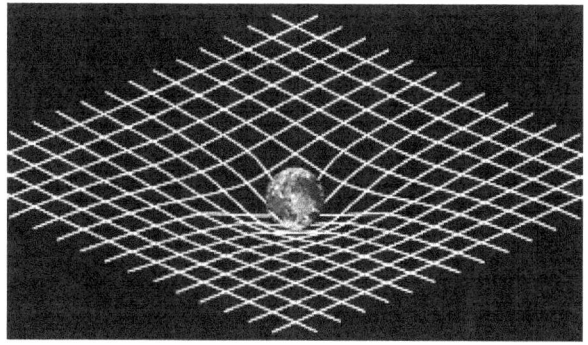

Fig 3.1: A demonstration on how objects bend the fabric of space and time.

Now have a look on that handkerchief and if you consider any one of the corner as a starting point (POINT A) for your interplanetary space travel and you want to go to the other side (POINT B). The general path you can take is represented in this diagram below:

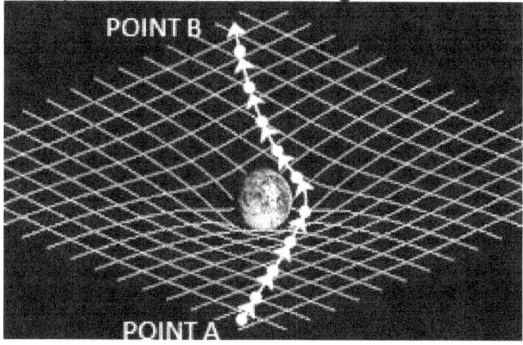

Fig 3.2: A general pathway of travel through the curvature of space and time.

Now understand what time is in this scenario, it is the total duration it will take for you to travel between point A and B. Now if the gravitational weight of the celestial object is huge, then the curvature will be much more, which means that the distance will be less and hence less time. Because a heavier object will bend the space and time curvature much more and this will lead to a shorter distance. A general overview of such a system is as follows:

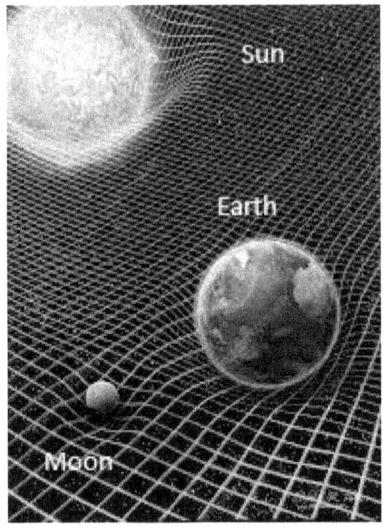

Fig. 3.3: A pictorial representation of how the mass of any object effect the curvature of space and time.

This effect is also observable for light because light is also affected by this curvature. Although this chapter is not directly related to explain how that is possible, we will learn it in chapter 7. However, a general representation is as follows where it shows how the gravitational field of a celestial object if sufficient large can bend the light.

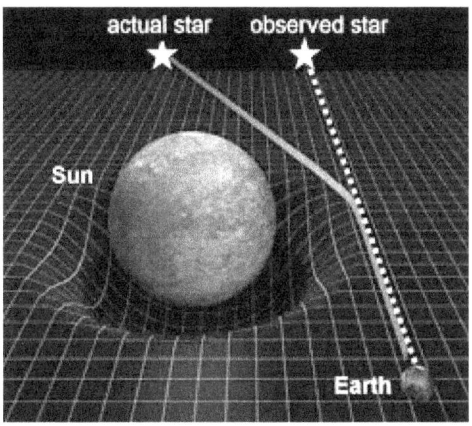

Fig 3.4: Bending of light under the influence of gravitational field.

Now I think you as the reader understand why time pass so slowly near neutron star or near black holes, because the gravitational field of these celestial objects are so large that it literally slow time by creating a large curvature on the space- time continuum.

But is this something we learned recently in the age of science of technology, you will surprise to know that it's not true. In many mythological tales such incidents are reported where it shows that time is passing slowly in different places. Let's have a look on few such tales, it will give you a new perspective regarding how earlier civilization tried to describe these observations.

The first tale is from the land of mysteries, India. Since Indian civilization is one of the oldest culture and it has shown time and again how in its ancient traditions and stories have hints to modern scientific theories and postulates. One such is the tale of Kakudami; he was a king in Hindu mythology. The tale goes that he had a daughter and he loves her so much that when she reached marriageable age, he wanted to get guidance from the Hindu god of creation, Brahma. To do so because he was very righteous and powerful he could travel to heaven to meet the god and return to earth. This trip took only few seconds for the king but when he returned to earth millenniums has passed. Now the story doesn't end here but for our understanding of time dilation we don't need any further to its plot. Now many of you might question mythical tales and that's your opinion. The ideas of these mythological tales are to shed some light on the scientific literacy that existed in earlier civilizations. Butbefore proceeding we have to understand what does time dilation means so let's have a look into it.

After understanding Einstein's theory of relativity we got the idea that time is not absolute, instead it depend on the reference frame from which it is measured. In his special theory of relativity, Albert Einstein gave this formula which relates how mass changes with increase in velocity.

$$M = M_0 / \sqrt{1 - \left(\frac{v}{c}\right)^2}$$

Where M is the final mass and M_0 is the rest mass.

But these are mathematical prospect of this theory; let's look into it from a more generalised idea in laymen term. Think in future we develop technology that can actually make travel with velocity faster than light which is known as FTL (faster than light) travel. Now if you board on such a space ship and started your interstellar journey even for decades, only few seconds will have passed for you but on earth decades will have gone. There is a beautiful Hollywood movie based on this concept.

But the question is you might be thinking that this is all theoretical. Well in fact this is happening in reality on the international space station which is rotating quite a high speed around earth. Although the speed is not nearly to the speed of light but the astronauts aboard the station actually moves few nano-seconds ahead of us, so in short they are kind of a time traveller. But can we use this concept and execute real time travel; well that's a topic for a later upcoming chapter.

The history of space and time

Let's have a look on how the relationship of space and time evolved throughout the pages of history. Over the course of development several notable mentions about these two were made by some of the greatest mind in science.

The first known work in the field of cosmology was done as early as 340 B.C. by Aristotle in his book on the heavens. In this book he mention how the earth is not actually flat but spherical in shape, based on its shadow during the lunar eclipse. A similar inference was made by an Indian astronomer and mathematician Aryabhatta. But the unfortunate thing is even today there are some people who are ignorant towards this basic fact.

With the passage of time in the year 1514 a model of our solar system was proposed by a Polish priest, Nicholas Copernicus. He published it anonymously in fear of heresy accusation by the church. In his model he

stated the sun was in the centre and all planets including the earth was rotating around it. This was one of the earliest mentions of the actual solar system but it took nearly a century after its publication to be actually even considered for reality. It was openly supported by two brilliant astronomer of the century, Kepler and Galileo and it came to know as the Copernicus theory. Although the model did not predict the actual elliptical orbits correctly, it laid the basic foundation of the science of cosmology.

In 1609, Galileo started his observation of planets by using his telescope which had just been invented. When he was observing Jupiter he found it was accompanied by several small satellites bodies (moon) which orbit around it. This discovery led to a new era of space exploration because it contradicts the idea which was postulated by Aristotle and Ptolemy. According to them, earth was stationary and every celestial object revolves around it. This discovery led to an age where space exploration was supported by actual science and not by superstition. At the same time Kepler modified his laws which even today established the basic idea of planetary motion. The laws are as follows:

Kepler's 1st law: Every planet orbits around its sun in elliptical orbits with the sun at its centre. This law shows how gravitational pull from the

star keeps the planet in orbit and in a unique way shows the important of gravity.

Kepler's 2nd law: The radius vector from the sun to a planet sweeps equal areas in equal times. This law also called as the law of area shows how the planet will always cover equal distance at every instance. This has a unique meaning behind its statement, this show that the time period of the planet depends on the orbit it is rotating and not on its mass.

Kepler's 3rd law: The ratio of the square of the period of revolution and the cube of the ellipse semi major axis is the same for all planets. This shows why each solar system has different rotation and revolution data because it depends on the gravitational pull of its sun. But as we will learn in later chapters that if a super massive neutron star or black hole is at the centre of a planetary system then the time will pass extremely slow for such planets. The most fascinating thing is even unknowingly Kepler showed the unique beauty between space and time. Here is the data which shows the pattern for our solar system.

Modern data (Wolfram Alpha Knowledgebase 2018)

Planet	Semi-major axis (AU)	Period (days)	(10⁻⁶ AU³/day²)
Mercury	0.38710	87.9693	7.496
Venus	0.72333	224.7008	7.496
Earth	1	365.2564	7.496
Mars	1.52366	686.9796	7.495
Jupiter	5.20336	4332.8201	7.504
Saturn	9.53707	10775.599	7.498
Uranus	19.1913	30687.153	7.506

| Neptune | 30.0690 | 60190.03 | 7.504 |

The detailed explanations of these laws were put forwarded by Sir Isaac Newton in his book Principia Mathematica Naturalis Causae, which was published in 1687. This was the first definitive discussion of the immense effect that gravity has on space and time. Before we precede any further, I would like to point one thing in the vast cosmos of space and time, gravity can actually pass through time. Which means it is not restricted by the concept of past, present and future; we will learn more about this interesting concept in chapter 8.

Now let's go back to the year 1691, this was the year when Newton got confused on the idea that if every object is having gravitational pull then why don't massive stars are not having any attraction between them. So he wrote a letter to Richard Bentley, another brilliant mind of that generation. Newton argued that since there are infinite numbers of stars in the known universe and they are spread over an infinite space. This was in fact an indirect realisation that the universe is expanding but it was too much to be understood at that time. A better understanding was reached by the mid of 20th century.

The expansion of space

One of the most significant events after the big bang was the expansion of space. As by now I guess you have understood that time is relative in nature, which was explained by Einstein's theory of relativity. This mean at the very beginning of the big bang, time itself was created. Later with this event the space started to expand with a speed greater than that of light itself. So now if someone asks you are there something faster than the speed of light, you know your answer. But how is this possible, well let's begin the journey to find it out.

As we today know that the universe is immensely large whose size is phantom to our imagination, then how can we assure that it is undergoing any further expansion. The journey to answer this question started in the early 1924, when Edwin Hubble first discovered other galaxies. Until that point it was thought that what we know as the Milky Way was the entire universe. But after this stunning discovery we realise how insignificant we are in the midst of cosmos. But how can we know how far away any celestial object is from us. You might have heard about the term light year, it's a measure of the distance that light will travel in one earth year time frame. We use this term to actually measure the distance between celestial objects but have you ever thought how scientists find these large distance. Let's have a look, but before that you have to know which parameter is taken under consideration while working out this distance.

The most important factor that remains constant across the universe is that there must be a star and planets in that system will rotate around that star. Now the brightness of any star depends on two factors, the luminosity of that star and how far away it is from us. So if we can calculate the luminosity of a star in another galaxy we can easily calculate the distance between us from its brightness. Hubble argued that certain type of stars will have a fixed luminosity based on the size and type of star. Hence if we can find such stars in any galaxy, we can predict it brightness and hence we can find the distance between us. If we perform such calculation to a number of stars in any galaxy and use its mean value we can get a fairly good estimate of the distance. Using this method Hubble calculated the distance of 9 different galaxies.

Have a look at this image below, it is showing the transition of a planet across its sun.

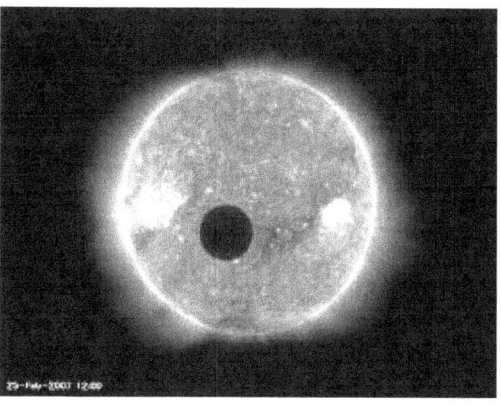

Fig 3.5: Transit of a planet in front of a star

When any planet transit across its sun there is a decrease in its luminosity and this vital information because it not only gives information regarding the star but also provides data regarding the revolution time and diameter of the planet along with its distance from its sun. But the stars are very far away from us, in most cases they are millions of light year ahead of us which means that the star we are seeing today is in fact a remnant of millions of years ago. But there is one thing that remains constant in all these stars no matter how far away they are, this is there light. Newton observed that the light from any star including our own sun when passed through a prism will undergo diffraction. This diffracted light will form a spectrum which is unique for each and every star but there relative brightness of each colour will remain same. But what does this information tells us, it tells us the core temperature of the star and what are the minerals present in that star based on the spectral analysis.

Now in the early 1900s, astronomers began their study to analyse the universe and the best path for them to do it was by studying the stars. What they observe was a shift in the wavelength of the light towards the infra-red part from the stars and these follow a trend from stars even in our own galaxy. The only answer to this phenomenon was Doppler Effect. This effect says that when a sound is moving towards us its frequency will increase and when it goes away from us its frequency will decrease. This can be observed in case of cars and trains as well; when we are on a platform and the train is approaching its sound will become shrill. Since frequency is inversely proportional to wavelength an increase in wavelength means decrease in frequency which corresponds to an increase in distance. Hubble continued in his life to research on these stars and galaxy and made an interesting discovery; the further away the galaxy is from us, the faster it is moving away from us. Honouring his life's work NASA launched a space telescope in 1990. The most fascinating thing is there are some data which suggests that the expansion is happening at a much faster rate than that of the speed of light. The tale is just getting started and we will learn more about this phenomena and how we can use it to explore into the final frontier of space.

This thrilling and ground breaking discovery that the universe is expanding marked new horizons in the field of space exploration. Both theoretical physicists and astronauts started working to find out what's there at the end of the universe and why universe keep on expanding. Unfortunately human technology has limit our potential at the present time to give definitive prove as to what this expansion actually looks like and why it is happening so. But we do have some hypothesis and with recent advances in physics and deep space exploration we are much closer to the answer than we were 100 years back.

The tale of universe and its beloved partner time is still pending. What we read in this chapter so far is just a small glimpse of the vastness of their glimpse. We are limited by the only thing we are trying to understand and that is time itself. Comparison to the age of the universe it is almost insignificant to the amount of time humans have been on this planet. We don't even know the history of our sister planets in our own solar system. But the one thing nature has given human is the perseverance and the quest to find the unknown and using this we have evolved from simple cave dwellers to space exploring species. In the next two chapters we will focus

on the legends of our universe and we will see how we can combine our ancient religions and science together to uncover some myths and know more about these legends.

Chapter 4

The birth of legends - 1

The study of space and time will be incomplete without the discussing the legends which make our universe so fascinating and who in their course of journey through time created the cosmos which today sustain life. And yes I mean life overall in the universe; the ingredients of basic life as we know is so common that there is life in the universe other than earth. It is not only foolish but completely illogical and unscientific to claim that earth only has life. The vastness of universe is so massive that we can't even presume the phantom of its vastness. But these two chapters are dedicated to those legendary celestial bodies which shaped the very space and time.

In these two chapters we will see how there is unique blend in science and religious tales. As I have always believed even as an atheist that science seeks to know the unknown but religion in its true form and not any superstition can provide with some essence to guide us in

our path. We have already seen one such story of time dilation in our previous chapter. So let's start this chapter by understanding what happened in the early days after the creation of the universe, which is immediately after the big bang.

THE FIRST MILLENNIUM

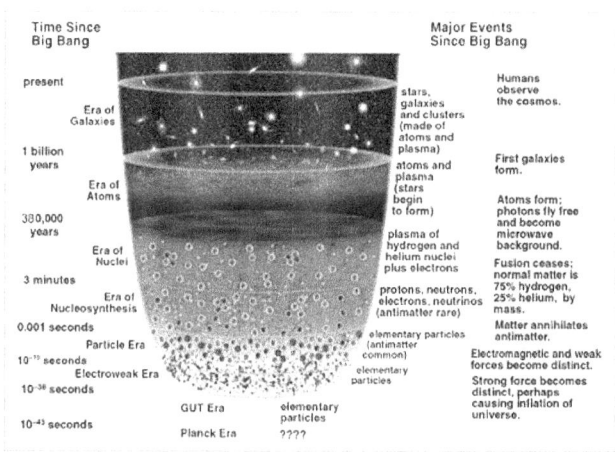

Fig 4.1: This shows the chronological events since the big bang.

The chronology of our modern universe begins at the exact moment when the big bang occurred. From a singularity spun open the entire new universe as we know today but without all the celestial objects. Was there another universe before our and is this a cycle of life? Maybe there lies a definitive answer but we can't make a conclusive statement because as we have seen in our last chapter that time was born with the big bang, hence it is practically impossible for our imagination to comprehend the criteria and conditions before the big bang.

That being said the CERN observatory had made some serious discoveries in the early life of the universe, several research works were conducted by two independent groups at UC Berkley, California which grant them the Nobel Prize in physics of 2011. But let's understand what occurred in the early universe after the big bang in simple layman's term. Initially think of the explosion of a water balloon, matter will have rushed outwards with explosion and started the expansion process thus increasing entropy. This, we have discussed in the earlier few chapters. As time passed by gravity develop and it starts to condense matter and this is where the story of our universe began and the tale of a beautiful cosmos started to unfold. This story is quite fascinating and in a way quite long because it involves several stages with the birth of legends. They further shaped the universe in their own unique ways and after 13.7 billion years today we are witnessing this grand canvas. So let's dive into this tale of time to witness the behemoths of the early universe.

THE NEBULA SYSTEM

Have you ever looked at the minerals around you like gold, platinum, silver and many more and thought where do they come from? Well they all originate where the first life enabling elements started to form, the giant nebula gas cloud systems.

Let us go back to 13.6 billion years ago after the big bang, there was a period of time when the current law of physics were not in effect, this is because the necessary forces and components of the universe responsible for these laws were non-existence. This time period is called the Planck Epoch. After few thousands of years passed by the core temperature of the universe started to fall and matter began to condense. Subatomic particles began to form, Higgs Bosons, and other particles were also formed during this period. After another few millions years, these particles forms the corresponding elements which we all see in our periodic table.

But you might be thinking how we know about the chronology of these events, none of us were there and there is definitely no time machine to go back in time. Well we have several data from Chandra X-ray observatory, Hubble space telescope and even some other deep space scans and they reveal the pattern of how nebula clouds are formed. There are some nebula clouds which are millions of light years away which means technically they may

not be there anymore because we are kind of looking into the past. Those who are unfamiliar with the concept of light year, it basically mean the distance light travels in one earth year time, so if an object is 1 million light year away we are basically observing a remnant of that object from the past 1 million years ago.

Fig- 4.2: The Pillar of creation, a nebula around 7000 light year away from earth. Courtesy of NASA

Nebula systems were one of the earliest features of the universe to be formed. They are gas clouds made up of hydrogen, helium and other ionised particles. Based on the element present they have their unique blend of colours. The reason nebula system is so important lies in the fact that they are the birth place of stars. In a way the cradle of life originated from them, because without stars we won't have the life sustaining minerals necessary for life and the universe as we know will not have existed.

THE LIFE OF A STAR

A star is one of the most fascinating celestial objects on the night sky. In every tradition and culture in some way or other stars are associated as an object of heavenly beauty. Humans have always been fascinated about stars but we never pay too much of attention to our closet star, our sun.

But the story of stars are not limited to just twinkling beauties and for legends, they are the source for life in this universe. The life cycle of a star depict the very essence of our universe and the cycle of life. Although stars have several roles which we will learn in chapter 6, they directly helped life to originate on our own planet but let's now focus on the cycle of the star.

Stars are created from nebula system which we just discussed. The entire cycle is described under a branch called stellar evolution. In recent years astrophysicists have invested a lot of information in observing such evolution in distant galaxies to get more insight of the early universe. Since we are discussing a story and in it time is the main protagonist, stars help a lot to know time better. The life time of a star can stretch for hundreds of millions of years, while the first humans evolved only 7 million years ago. The fascinating way by which scientists observe stellar evolution is due to the speed of light which create a natural time lapse. This is because the speed of light is 3×10^8 m/s, and

the term one light year is used to express the distance travelled by light in one earth year time. Hence when any star is formed galactic distance apart light even with its incredible speed will take a very long time which is in millions of years to reach earth. This allows scientists to observe the entire time period of a star's life. In fact all of us are observing this time lapse every time we look on to our night sky. So next time you observe a star consider the fact that you are actually observing a past scene and you are literally viewing through time, in a sense. This particular scientific phenomenon has its own drawback as well, when we say why we can't observe any habitable planet and aliens in the universe the reason is here. Just imagine if an alien is looking towards planet earth from a distance of 50 or 100 million light years away, for that alien earth will still be an uninhabitable planet because the alien will be observing from the past.

Now coming back to the birth of star; when dense gas clouds of nebula condense under extreme pressure. Such clouds are often called as nebulae or molecular clouds. Once condensation starts, self-sustaining nuclear reaction helps to keep the core active and depending on the mass of the star can last for hundreds of billions of years. Some stars systems are discovered frequently which are older than our current universe may be a remnant of past universes.

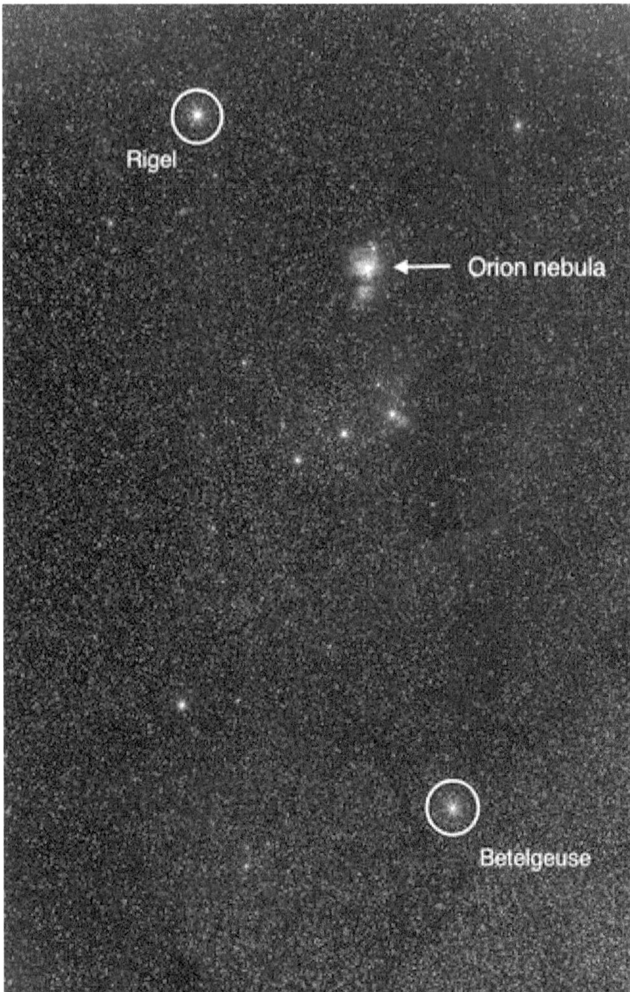

A picture of the Orion constellation with Betelgeuse and Rigel who are one of the largest known stars in the known universe

Initially while the gas clouds starts to condense they are highly unstable. The initial star phase is called a proto-star and over time it grows to the actual star also known as main-sequence star.

Almost all known star in the known universe sustain them using nuclear fusion where two hydrogen nuclei combined to give one helium. Now depending on the mass of the star other elements can be synthesised. For example our sun cannot synthesis heavy metals like lead, iron or noble metals like gold, platinum etc. These elements came from much larger stars and depending on the distance of these stars the concentration of the elements varies in our earth's crust.

Once the life time of the star is over, which means when they can no longer sustain fission, they will start expansion. In this stage it's called red giant. The star become massive and expands until it collapses under its own weight. If it has sufficient enough mass then it can form black hole else it will form white dwarfs. There are other pathways it can follow, it can become a supernova or pulsar or neutron star. We will discuss them and their

role in shaping the universe in the next chapter.

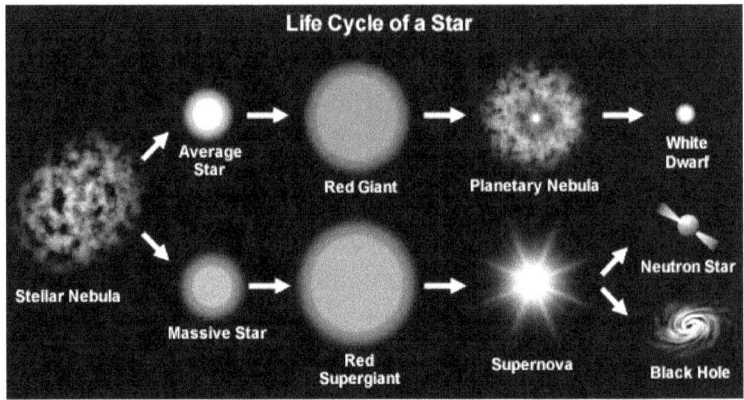

Credit: NASA

Black Hole

If there is anything which can be given the title of the most mysterious object in the known universe, then no doubt it will be a black hole. It can't be possible to discuss about time without mentioning about black holes and its role. The prediction of black hole was done in Einstein's theory of relativity but it took quite long to actually prove its existence. In fact the breakthrough work to understand them is so recent that the noble prize of 2020 was awarded for its research to three scientists, Sir

Roger Penrose, Reinhard Genzel and Andrea Ghez.

In simple layman's term a black hole is a disturbance in the fabric of space and time where gravity is so strong that even light particles (photons) cannot explain that. To keep the details fairly simple we will not indulge ourselves into complicated mathematics of black holes but have an overall outlook of this topic. But before we understand the mysteries behind black holes, let's understand how they are born.

Black hole is one of the ways the death of a star could go; other ways are supernova, pulsar and neutron star. The formation of black hole and its time period of existence are two different things. Any object with a sufficiently large mass can be converted to a black hole if the mass is compressed in an extremely small sphere. This idea was worked upon by Dr. Stephen Hawking and the quantity that defines such formation is called as Hawking radiation. Hawking's radiation is a type of black body radiation which is given by black holes. Any such black hole will absorb all lights from the entire spectrum including UV and infra-red.

The first work over the concept of black hole using Einstein's general theory of relativity was done by **Karl Schwarzschild.** He predicted that depending on the amount of matter present in the star, the matter has to be compressed to a particular size. The radius of this compressed state of matter is called as the **Schwarzschild radius.** In common term this is also known as the event horizon. Once any object enters the event horizon of a black hole, it's physically impossible to escape its gravitational pull. This radius is depended on the mass of the black hole and hence can be defined as follows:

$$R_s = \frac{2GM}{c^2}$$

Where; R_s is the Schwarzschild radius, M is its mass and G is universal gravitational constant.

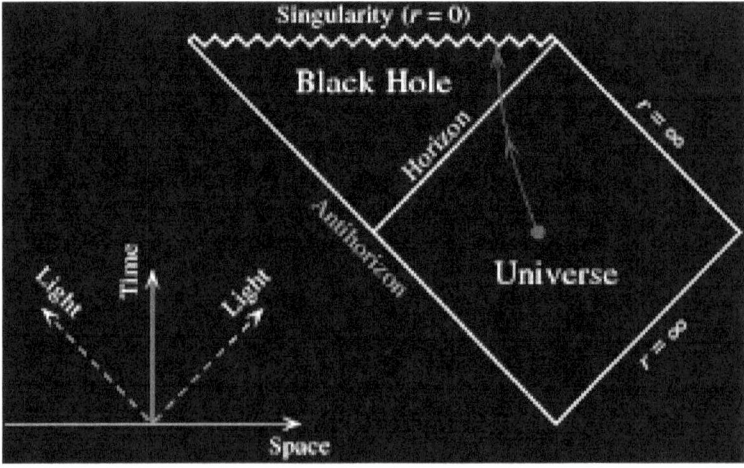

Penrose Diagram depicting black hole in space time continuum. Designed by Roger Penrose. Credit: University of Columbia.

From the above formula it is clear that the radius of the black hole is independent from its spin or its electric charge. In reality they are the objects which define gravity to us and in a way showed us the true meaning of Einstein's general theory of relativity.

There are several mysteries of the universe yet to be discovered but in the late 20th century two of the most brilliant minds of the time worked together over this concept and tried to device the origin of universe from Einstein's relativity. These brilliant minds were Dr. Stephen Hawking and Dr. Roger Penrose from University of Oxford. Later Dr. Penrose won the Nobel Prize for physics of the year 2020.

Dr. Stephen Hawking with Dr. Roger Penrose during their collaboration work on cosmology

The concepts behind black holes are quite extensive and we won't venture into it, else we will be lost into its vast knowledge. In simple term we can say that near the event horizon which marks the Schwarzschild radius for the black hole, time moves extremely slow relative to that of earth or any other similar planet in the known universe. However as we cross this event horizon and enter the singularity, which is the centre of the black hole time stands still. This is a basic difference between Einstein's general theory and Newton's idea of motion in case of Einstein's time is relative while for Newton it's absolute.

Before we proceed to the structure of black hole, it is important to mention two unsung heroes in the field of astrophysics whose work in a way contributed to their collaboration of Hawking and Penrose. These two among many others are AK Raychaudhuri and CV Vishveshwara. Both of them published ground breaking research with mathematical models which helped to create the foundation work on which later models of black holes are designed. In this next section we will just understand the theoretical ideology with simplified mathematical aspect behind the formation and structure of black hole.

Structure of black hole and explanation for its formation.

Have you ever imagined how theoretical physicists predicted that the shape of black hole will be spherical years before we actually obtained its picture? The answer lies in basic geometry; sphere has unique geometrical benefit one of which is distribution of sheer load over its entire surface equally. This is the reason why water droplets take this shape and why celestial objects like stars and planets are frequently of this shape. We will learn more about them in the next chapter.

According to Newtonian theory, a particle can execute circulatory motion from any arbitrary position from the central force. This however is not true in case of Einstein's general theory of relativity. As per this theory black holes have a stable radius, innermost stable circular orbit (often called the ISCO) which depend on its mass and Schwarzschild radius (spin zero),

$$R_{ISCO} = 3R_S = \frac{6GM}{c^2}$$

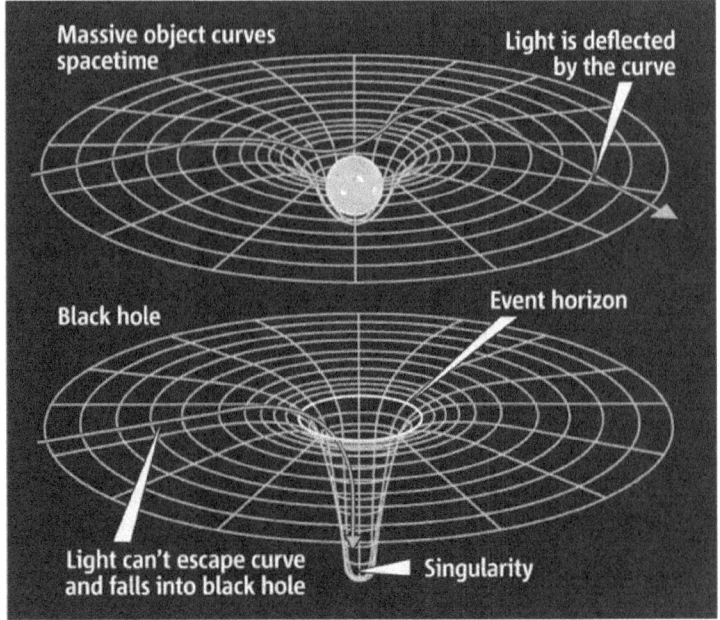

A schematic representation of black hole as per the general theory of relativity.

Due to its immense density, black holes can curve space and time as shown in the above picture. These massive objects are the perfect example of law breaking scenario in physics. Time is not the only thing that is affected by gravity; light can change colour simply by being put under a gravitational force. To escape a depression in space-time, light must lose energy, but Einstein had previously shown that light travels at a constant speed in special relativity. How can light lose energy without losing speed? Well, wavelength and energy and inversely related, meaning by increasing one of these variables, the other decreases. For light to lose energy, it simply must increase its wavelength. This causes a redshift in colour that affects all light travelling through valleys in space-time.

In reality the picture of black hole which we have recently captured shows the light from the event horizon and not from the actual singularity. Because nothing can escape the gravitational pull of black hole. It is very vital to understand that there is no evidence that black hole will lead to another universe in the multiverse, which is pure science fiction.

Now not all black holes are gigantic in size and those which are smaller in size are extremely dangerous to venture forth. This is because the moment any object enters into such a black hole, it will undergo a process called **sphagetification**. According to this process the gravitational forces of the black hole will rip apart anybody because the forces experienced by the lower half of the body will be much higher than that experienced by the upper part. The only black holes humans can venture into are the gigantic ones whose gravitational anomaly is evenly distributed. Kip Throne designed one such black hole with theoretical concept which he used for Christopher Nolan's

movie Interstellar. It was so scientifically accurate that it later matched quite a lot with the real picture taken later by the team of scientist from MIT, USA.

Another form of black hole is a **quasar** in which a super massive black hole with an extremely gigantic mass, few millions of solar masses is surrounded by a gaseous rotating disc. They are extremely bright in nature and the gas is from the dying star. Most of the time it is very hard to differentiate a quasar from a star but they play an important role. They are living proof of our distance past when these mega stars were actually there in our universe. Recently a quasar was discovered which was created just 700 million years after the big bang is. Such massive objects tell the tales of our past; it's just needed that we listen to it.

The Black Hole Information Paradox

Within the realm a black hole, the physics of general relativity and quantum mechanics, the science of subatomic particles, come together and are both applicable. This may not seem terrible, as both theories are widely accepted and understood, but it is worse than expected. In reality, general relativity and quantum mechanics are completely incompatible, and they both predict different things to happen in black holes, causing paradoxes to form.

The main paradox created is the black hole information paradox. This paradox was created by a discovery made by Stephen Hawking in 1975. In this year, Hawking postulated that black holes radiate particles, shrink, and eventually disappear from existence. This phenomenon is known as Hawking radiation, and it changed the way physicists looked at the big, bad black hole.

To theorize Hawking radiation, Stephen Hawking used one of the strangest ideas from particle physics: particles and their antiparticle pair break the law of conservation of energy and form from nothing, destroy each other, and end up keeping the law of conservation of energy intact. The creation of a particle and their antiparticle pair is very common, and it occurs nearly everywhere. This includes near the black hole's event horizon. According to Hawking, a particle and antiparticle pair formed near the event horizon has the possibility to have one of the particles sucked into the black hole. If this occurs, then the particle's pair will escape the black hole to infinity. Because the particle and antiparticle do not destroy each other, the particle created breaks the law of conservation of energy by simply existing and moving. To combat this, something must lose energy. To conserve energy, the only thing available to lose energy is the black hole itself.

Because energy and mass are equivalent according to Einstein's, the black hole also loses some of its mass. This creates an image of the black hole radiating particles and slowly evaporating away until nothing left exists.

Stephen Hawking's theory posed serious problem for physics. On one hand, general relativity states that nothing can escape a black hole and information will be lost forever. On the other hand, quantum mechanics states that information must be conserved.

When Hawking radiation was first theorized, quantum mechanics faced serious trouble. Because the particles do not seem to come from inside the black hole, information from the black hole apparently disappears and violates the Law of Conservation of Information, a critical law to quantum mechanics. To some physicists like Stephen Hawking, the simple conclusion was that information could be destroyed. This conclusion was troubling for some physicists because it would mean quantum mechanics was incomplete, but they eventually came up with theories to combat the paradox.

The main theory to combat the black hole information paradox and Hawking's conclusion that information can be destroyed was created by Leonard Susskind of Stanford University. He believed the opposite of Hawking; quantum mechanics are correct and general relativity

needs to be fixed. Susskind postulated the new and theoretical idea called black hole complementarily. It states information is both trapped inside the event horizon and reflected on the surface of the black hole. Observers inside the black hole only see the information inside the black hole while observers outside it see information gather up at the event horizon and radiate as Hawking radiation. Because neither side can communicate with each other, there is no paradox created. This theory requires another idea to be true, holography. This idea says that the physics in the three-dimensional black hole must be describable as physics from the two-dimensional surface area of the black hole. While black hole complementarily and the resulting holography seem insane, the ideas gained popularity quickly. Eventually, the concept that general relativity needed to be fixed while quantum mechanics was right became the majority in the world of science. Eventually, even Hawking sided with quantum mechanics when he redid his calculations and came to the conclusion that as an object falls past the event horizon, it disturbs the black hole's radiation field.

Looking Forward

The basic idea which you have obtained till this point will help you to understand the later part of time and space and how theoretical physics is changing our concept of the universe. As we proceed we will learn more about them and how can we perform time travel, even in a purely theoretical state. However a very important aspect of science is the idea of discussion. It is to be noted that science is not build by the thinking of one person but with the shared knowledge of everyone over the generations. There was a time when nuclear fusion seemed impossible, a time when the greatest minds in the world use to think that atoms are indivisible. But with the advent of modern technology we have proved many such things which were once just science fiction.

With that being said lets go the second part of this chapter to understand some other celestial objects in the universe and their formation.

Chapter 5

The birth of legends -2

Pulsars

They are highly compact magnetised stars which emits beam of electromagnetic radiation from its poles. Pulsars are actually a type of neutron star but there is a minute difference, all pulsars are neutron stars but all neutron stars are not pulsar.

In the previous chapter we have discussed on how a neutron star is created in the life cycle. When the star is about to die there are many ways the path can move including that of a neutron star. They are formed when a star is compressed to form a supernova, they retain their high angular momentum and that gives the pulsar their extreme high rotation speed. They have charged components which helps them to emit electromagnetic beams. These are like light houses in the grand cosmos and can be used to obtain valuable data about distances in the universe and also about the age of a given solar system.

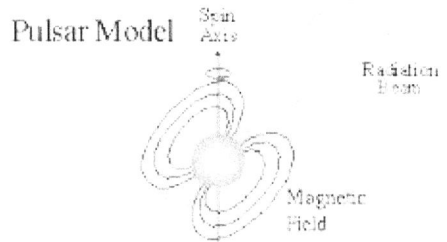

Courtesy: Ohio State University

Fig 5.1: Pulsar model

The first pulsar was detected back in 1967 by Jocelyn Belland and Antony Hewish (Nobel Laureates) by monitoring small radio frequency.

Jocelyn Belland and Antony Hewish

The idea of neutron stars was first proposed by Walter Baade and Fritz Zwicky in 1934, when they argued that a small, dense star consisting primarily of neutrons would result from a supernova.

Pulsars in close binary systems are also powerful energy sources. Consider dropping 1 gram of matter (about the mass of a paper clip) onto a neutron star from a great height. When the mass goes `splat' onto the neutron star, 30 trillion joules of energy are emitted. That's 40 times the amount of energy as one would get by fusing 1 gram of hydrogen into helium (200 million times as much energy as one would get by burning a gram of hydrogen). Thus, dropping matter onto a neutron star is an energy source even more efficient than nuclear fusion or fission. The matter that goes `splat' is heated and emits X-rays.

But why do we study them. The reason lies in the fact that pulsar tells major tales of the universe. They show the true nature of matter and in a way shows the fundamentals of x-rays from a natural source. Some of their applications are as follows:

- ❖ Since they are called the light houses of the universe they truly act in navigation. While drawing star charts they are very helpful in actually identifying various systems in the universe.

- ❖ The emission done by pulsar is very precise and in fact shows the fundamental nature of time itself. They are used as a reference point for atomic clocks used in our satellites and international space station.
- ❖ Interstellar medium is a term which is used to defined ionised, molecular and atomic gases which exist in the universe. The pulsar beam is so powerful that it can be used as probe to study these ISM.
- ❖ Pulsars are excellent in understanding space and time. When a pulsar exists within the curvature of space and time near a super massive black hole, their pulses can be used to study gravitational waves.

A very vital point to understand is that pulsars are only detected when their electromagnetic beam is pointing towards earth because we know from basic school knowledge that light can't bend and so is for the beam of pulsar.

Dedicated research on pulsar is still going on with special efforts from NASA and other space agencies to explore the depth of cosmos and to unravel the mysteries of deep space.

Neutron Star

Neutron stars are also one of the stages which a star passes by during its life cycle. They are formed from supernova while gaining an exponential factor of density in comparison to the original star. The neutron star may develop into a pulsar or may just remain a neutron star in perpetuity. It is to be noted that a neutron star will never die and with time due to release of radiation, it will cool down but will never die. Other than a black hole, neutron stars are some of the densest stellar object ever found in the universe.

If you are familiar with the term angular momentum which you may have learned in your middle school, you should know that it plays a vital role in neutron star. Once they are formed they keep on rotating but they stop generating heat. The neutron star has a different state of matter which is a highly dense state of Fermion particles. Pauli Exclusion Principle* explains why their further collapse is prevented. Don't worry we will discuss each of these things in details but before doing so we should acknowledge a great scientist, a Nobel laureate who worked with neutron stars. His name was Subrahmanyan Chandrashekar; the nephew of the famous Nobel laureate, C.V.Raman.

Chandrashekar worked on these cosmic bodies and discovered what is called as the ***Chandrashekar Limit***. This limit defines the maximum mass possible for a stable white dwarf. This limit is greatly helpful to understand a stable neutron star as well. We will not go into much deeper of this topic because this falls under the domain of astrophysics. Neutron stars have a degeneracy level among them which means all of them are in a similar energy level, but neutron degeneracy is different from electron degeneracy. But again this falls under quantum theory and it will be a very lengthy topic to discuss. In conclusion a neutron star is about 10 kilometre (approximately 6.2 miles) and with a mass of 1.4 solar mass (1 solar mass = 1.98×10^{30} Kg).

S.Chandrashekar, Courtesy of Smithsonian Institute, USA

But neutron stars are extremely far away from earth and the only way we can observe them is by studying their heat signature. Some of them are thousands of Kelvins in temperature and with extremely high gravitational and magnetic field. Due to such extreme gravitational anomaly, time flow differently when near such dense objects.

It is to be understood that angular momentum is conserved in these stars and hence the centre is rotating at a much faster speed than their external circumference. To get a perspective, the escape velocity of planet earth is about 11 km/sec and that of neutron star is about 150,000 km/sec.

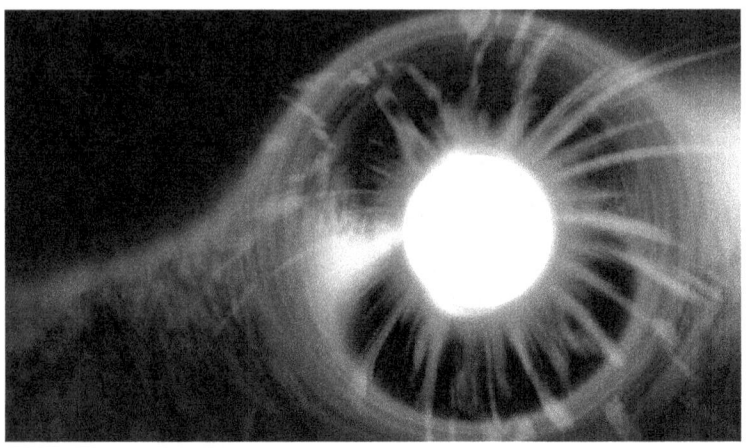

A typical neutron star

They are like mythical creatures, born out of the depth of a dying star from the heat of a supernova, a true legendary phoenix. But they are real and there are thousands of such neutron stars discovered in our own Milky Way. But a word of caution these celestial beasts have such a high density that just an impact with them will render any matter to its atomic particles, so we are not going to any neutron star in the foreseeable future. In one word they are truly a celestial phoenix.

> - **Pauli Exclusion Principle**
>
> This Principle states that in any given quantum state two fermions particles cannot have the same four quantum numbers. This is also applicable in case of electrons where it is impossible for two electrons even of the same atom to have all four quantum numbers same.
>
> This is just a summary, to apprehend it better you need to learn about the four quantum numbers and quantum description of atom.

Why to study Neutron star?

The tale of these legends won't end if we keep on studying in details because countless research works are being performed and numerous research papers are being published in reputed peer-reviewed journals over them.

But why do we need to study them and how is it related to the main aim of this book which is time. To understand this we have to know a phenomenon known as **Gravitational Lensing.**

If you remember your basic middle school class science you should know that light never bends from its original path which means if you are standing behind a wall, another person standing on the opposite end cannot see you, but that is the beauty of these heavy objects

that they can alter the path of light. When any such dense objects come in between the source of the light and the observer they can bend them. This effect helps us to even observe distant celestial objects even if they are blocked by other celestial objects. The image of the neutron star shown there is a typical example of gravitational lensing.

The deflection of light is given by the formula:

$$\theta = \frac{4GM}{rc^2}$$

Where;

θ = Angle of Deviation

r = Radius of the neutron star or any other dense material

c = speed of light

M = Mass of the heavy object.

G = Universal gravitational constant.

What does the idea of gravitational lensing proves? It shows that gravity can affect light on a practical scale. And since light is closely related to time, gravity is the only force which can pass through the barrier of space and time. To understand this logic on a cosmic scale, it means the only way we can interact with our past is with the help of gravity. But that is a discussion of a later chapter.

Chapter 6

A humble Journey

Carl Sagan in his famous book, "The Pale Blue Dot" describes astronomy as a humbling subject as it teaches human beings the true meaning of life and the spirit of humanity. We have always considered our life to be very simple but in the grand cosmos for us to have a life a lot of things have to be precisely move in a given order.

Every religion in this world has some works regarding this event and in their true form depicts humility and universal brotherhood. Looking from a scientific perspective it is vital to note that these events make each of us unique. Whenever you feel upset about anything in life, just remember these events as they are what make you unique. So let's dive into the past to have a glimpse of our humble journey because at the end it's the journey which matter more than the destination.

In the previous chapters we have understood the scenario after the big bang and some of the most glorious celestial bodies which were formed after the event. But the most interesting part in the entire universe is yet to happen. After the big bang the new universe immediately started to expand and this is already explained earlier under the concept of entropy. But along with this expansion there was another crucial factors and that is the star systems. For a moment let us not focus on our

solar system in particular but to the general idea of a star system (*Not to be confused with star cluster*).

A star formation takes millions of years since the beginning of the cluster condensation of the nebula gases. Once a star or binary stars are created they will generate their own gravitational field which is sufficient enough to engulf quite a large part of space.

Any celestial bodies within this part will fall under the influence of this star system. The reason lies in the fact that for any object to remain in a uniform circular motion the basic laws of physics have to be obeyed. This is applicable to every star system across the universe. The law is simple that the centripetal force should counterbalance the centrifugal force.

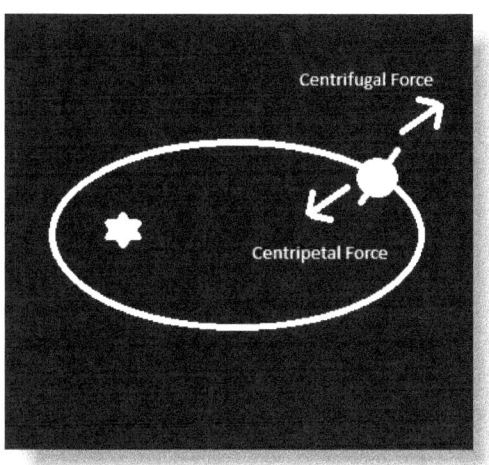

The forces keeping the planet in a uniform circular motion

The planets will follow an elliptical path and they are governed by Kepler's Law which are as follows:

1. All planets revolve around the sun in elliptical orbit keeping the sun in one of the foci.
2. A radius vector joining any planet to the Sun sweeps out equal areas in equal lengths of time.
3. The square of any planet's orbital period is directly proportional to the cube of the length of the semi-major axis of its orbit.

These laws are applicable to every planet in other star systems as well. But the most fascinating type is a twin star system and more fascinatingly triple or higher star system. The question is if it was good if we had a twin star system or what are the events in the timeline which helped create our solar system.

This chapter will focus primary on our solar system because if we want to understand time in general and to understand the exploration of space and become an intergalactic species, we have to understand our very own solar system and the details about its formation.

Planet Formation

The science of planet formation is really interesting and the best part about it is our ancestors in the early ages understood its mysteries. Great personalities like Galileo, Aristotle, Aryabhatt and many others understood that our earth is oval (More precisely Geoid Shape). However their time was very unforgiving and religious sentiments took the form of bigotry and the rest is history to all of us.

Fortunately we are living in an age of science and technology and we can understand the idea of planet formation because this idea will help us understand that is it possible to

transform harsh planet like mars into habitable ones. So let us understand how any planet is formed:

- ❖ The most common process of planet formation is condensation of molecular gas clouds. In this process the gas clouds near a star system will start to condense under the gravitational influence of the star and thereby form corresponding planets. This is part of the nebula hypothesis and generally starts at the very early stages after the stellar birth from a nebula. This is to be considered as a very common way by which most planets are formed across the universe.

- ❖ But planet formation appears to have happened relatively rapidly. Small bits of dust and gas began to clump together. The young Sun pushed much of the gas out to the outer Solar System and its heat evaporated any ice that was nearby. Over time, this left rockier planets closer to the Sun and gas giants that were further away. This hypothesis was first postulated by Pierre-Simon Laplace, who was a famous mathematician and astronomer who gave us the Laplace Transform.
- ❖ Another important factor in planet formation is angular momentum. Just picture our own solar system and its 8

planets (I know we all love Pluto but what to do). Now what is a unique property that you observe? It's the size and their angular rotation speed. The inner planets are all smaller in size in comparison to the gas giants. This is because as the planets began to condense the change in angular momentum affects their relative position to the sun. Planets closer to sun will have a faster rotation then those gas giants. This is why the length of day is different for different planets and same for the length of year. So we can see how time is relative even in our own solar system.

❖ But there is another way of planet or dwarf planet formation which is by collision or impact formation. This happens when two celestial bodies collide and that give us a series of objects ranging from asteroid fields and dwarf planets to natural satellites. This is why planets like Jupiter and Saturn has so many natural satellites.

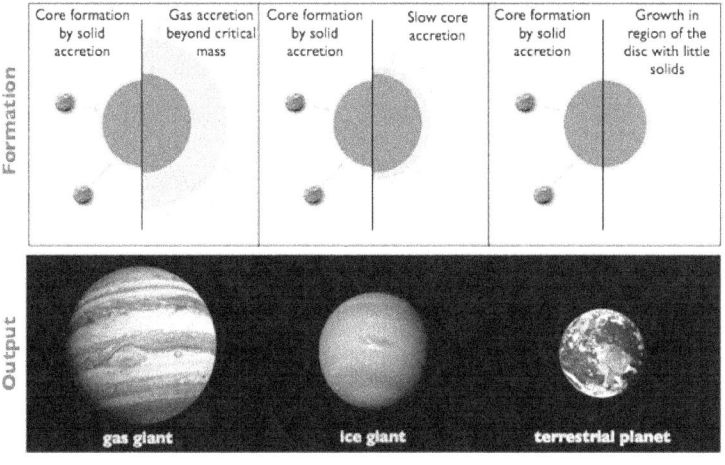

The dust around a star is critical to forming celestial objects around it. Dust around stars contains elements such as carbon and iron which can help form planetary systems. When a star is in its forming disk, otherwise known as the *T Tauri* phase, it is ejecting extremely hot winds dominated by positively charged particles called protons and neutral helium atoms. Although much of the material from the disk is still falling on the star, small groups of lucky dust particles are crashing into one another, clumping into larger objects. Dust clumps become pebbles, pebbles become larger rocks that grind together to expand. The presence of gas helps particles of solid material stick together. Some break apart, but others hold on. These are the building blocks of planets, sometimes called "planetesimals."

Even after the initial stages of planet formation, there are several important steps before the mass of rocks and metals can be termed as a part of planetary formation. Have you ever considered why our earth is so unique to life formation?

Although there is a dedicated chapter in solving this mystery along with the mystery if we are alone in the universe, we have to learn about one fundamental idea which is our core formation and magnetic field.

CORE & MAGNETIC FIELD

During the initial stages of planetary formation the heavy elements mostly iron and nickel sink to the bottom of the planet. They form the liquid core which is free to move inside the mantle as per the spin of the earth along its axis. This phenomenon creates a magnetic shielding effect which protects earth from harmful solar radiations and other cosmic rays from space. A pictorial representation is given below.

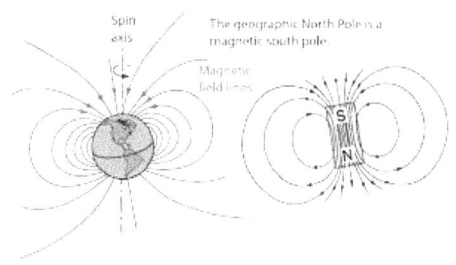

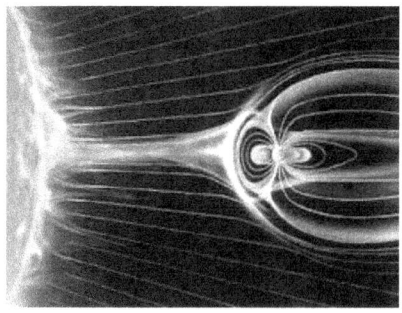

Pictorial representation of magnetic shielding effect of planet earth

This magnetic effect is also responsible for the magnetic poles of our earth which have helped countless human expedition possible using the magnetic compass. Other than its several applications this magnetic field plays a crucial role in maintaining the integrity of our atmosphere and sustaining life in our planet.

Sometime even something very small becomes the most significant factor in something so beautiful. Our neighbouring planet mars doesn't have this magnetic shield which is why it will be extremely dangerous to venture as life sustainable as we have on planet earth.

The Oort cloud

If you are asked about our solar system you will picture something you saw most of the time in books or on some magazine cover and that's not at all the reality.

The solar system in reality has a giant cloud of debris consisting of asteroids, dwarf planets and celestial wastes collectively called as the Oort cloud. A pictorial representation of the Oort cloud is given below:

The Oort cloud at the edge of solar system

This region is surrounding the Kuiper belt region and has played a very significant role in bringing life on planet earth.

Just imagine the entire water on our planet including glaciers, underground waters, sea, rivers, oceans etc. combined together. Do you think all this water was created during the chemical synthesis by chemosynthetic bacteria?

The answer is no, a popular hypothesis is that a huge part of this water including the heavy water (D2O) was brought to earth by asteroids from the Oort cloud. So in a way every aspect of solar system worked together in a series of events to bring our beautiful life supporting planet.

Science is a beautiful self-reflecting subject, in its true form it makes you more spiritual and humble then you can imagine. Sometime in our life we life in a grandiose sense and many time in the history of mankind such evil people have committed horrible crimes to fulfil their ego. Knowing our humble beginning brings us closer to our own self and helps us focus on the important aspect of our life.

Our planet formation shows that beautiful things takes time and perseverance, but most important grace. We often consider science something as anti-natural but science is the

only subject which helps you to understand nature from a very non-judgmental point of view.

At this point we have reached half way to our journey of understanding space and time and how life came into existence. It is so fascinating to thing somewhere a minuscule change would have drastic effect and maybe we would have never been standing here living our life.

We have several happy moments and some really difficult moments in our life and sometime it may affects us in a negative way. But all things considered just wait for the right time. As we have seen time is the ultimate driving force in our universe and at the end it will make all things right back to its actual track as it was supposed to be.

So as we proceed to our next chapter, we will have a broader look on how time played a very crucial role from earth's perspective and also is there life on other planets and if so why can't we contact them. In this next chapter we will understand the role time played in our creation and the concept of free will and are we the master of our own will.

Chapter 7

Time - A blessing

Humans from time beyond history have always tried to control our fate. We call ourselves the master of our own path and that is what true motivation and hope is. This is what has made humans from cave dwellers to space exploring species. But if you have read any religious texts we have always been taught to be humble as our choices are not our own. So which part of this statement is true?

Luckily science draws its answer from logic and experimental analysis, so by the end of this chapter you will learn several scientific facts about time and its role in bringing life along with the fact is our destiny already fixed? No matter what conclusion you get at the end, I can assure one thing you will see life and every opportunity from a new perspective.

Before beginning the chapter I would like to quote two quotes which truly describe the spirit of this chapter,

*"A man who dares to waste one hour of time has not discovered the value of **life**."*

*"What you learn from a **life** in **science** is the vastness of our ignorance." – David Eagleman.*

The key to a happy life is accepting that every day is a new opportunity. To thank for what we have in life, to strive for what we want but at the same time to have grace and humility in our heart. We always think that philosophy and science are different, in reality they are the opposite side of the same coin. This chapter will actually put the true perspective of life before the readers in a scientific approach and how because of these lucky events in their perfect synchronization we are here on our beautiful planet.

Formation of Earth

In our previous chapter we have discussed about the formation of planets in general and few key points about our earth. Here we will have an in-depth review of the various events which ultimately favoured our creation and how time plays an important role in it.

1. Position of Earth

The earth is located at a perfect location with reference to our sun. This zone is known as the goldilocks zone in laymen's term, astronomically this is known as the circumstellar habitable zone.

For life to sustain, several organic reactions have to work in perfect symphony. For this a very important factor is temperature, if a planet is too close to its star it will be extremely hot to be survivable for any life and if it is very far then due to extreme cold condition, life will cease to exist. As we know any planet's position relatively depend not only on its time of formation but at the same time it depend on the mass and rotation of the planet, which we discuss in the next point.

The formation of the earth was in a perfect time in order of creation of other planets of our solar system; else the earth could have become a dwarf planet thereby crossing eclipse of moving into a bigger revolutionary axis. If such conditions existed then life as we know won't be there. It is important to understand that without proper gravitational balance, our planet could have become a rogue planet. These are planets which do not orbit around any star, instead keep surfing the entire universe. Such planets cannot sustain life and can even undergo collision with denser celestial objects like supernova, neutron star etc.

2. <u>**Mass, rotation and revolution of our planet**</u>

The exact mass, rotation and revolution of our planet play a significant role in the position of our planet. As we know angular momentum depends upon mass and radius of the object. If the earth were bigger in size or heavier in mass or vice versa our entire orbit around the sun would be different. This means our precious goldilocks zone would be lost and the concept of life wouldn't exist in the first place.

But how does time plays a role in it?

Well the answer lies to the formation of Jupiter, Saturn and our moon at the right time. Asteroids, especially giant ones can at times increase the core mass of the planet and if regular hits are taking place then the entire orbit may get altered. Luckily the formation of Jupiter prevented any massive hits and at the same time prevented regular hits from life eradicating collisions by such asteroids. As we know one such asteroid hit eradicated the dinosaurs. The role of Saturn is much more interesting and we will learn about it in a later point.

Similarly the formation of moon has very significant impact over life on planet earth. Not only in the initial days of formation of life but moon plays a very significant impact in our daily lives. This is why is almost every ancient culture has a special place or myths on the

moon. The formation and importance of moon is described in the next point.

3. Theia Impact

Based on the study of lunar rocks and several scientific researches a hypothesis was made which is known as Giant-Impact hypothesis or Theia Impact. It suggested that moon was formed by an impact of a mar size planet and earth.

It was first suggested in 1898 by George Darwin who using centrifugal forces and Newtonian mechanics proved that the orbital path and angular rotation of both earth and moon are similar. This idea was challenged in 1946 by a professor of Harvard University, Reginald Aldworth Daly. However these discussions remained dormant for many years until it got it recognition. Later it was also worked by Canadian astronomer Alastair G. W. Cameron and American astronomer William R. Ward. The most significant result was published in 2016 based on an extensive study of the lunar rocks.

As per the latest findings, approximately 4.5 billion years ago (about 100 million years after formation of solar system) during the Haedon

era, when earth was in its initial stage in a burning state the collision took place.

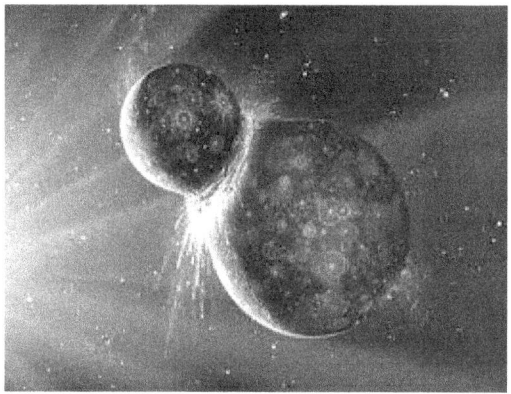

Artistic rendering of Haedon era

The lunar surface shows that the density of moon is in fact a combine effect of both earth and that planet which collided with earth. Due to this relatively sufficient dense surface with a close proximity to earth, moon has always affected tidal movement and in turn affected the weather and local conditions of many places on earth.

Here also time plays a very important fact, the collision happened when the mantel and crust of earth was not completely solidified. If the collision happened in any time after the complete crust is intact then we could have lost entire planet into space debris. In a way till now it can be observed that how the perfect time is a crucial factor besides having the right ingredients in creating life.

4. Earth's magnetic field

We have discussed how important the earth's magnetic field is. It is the reason we are having a stable atmosphere and it acts as a shield to protect us from harmful UV radiations, cosmic radiations and solar storms.

The reason for our magnetic field is already discussed but the time of its creation is very important. If the asteroids and comets bringing iron from the depth of the universe hit earth little late after its mantle has cooled down, then those hits would have destroyed our life-sustaining planet.

5. The uniqueness of gravity

Gravity is a mystery in itself and in the next chapter we have a dedicated discussion on it. However in this point we will discuss how gravity in general helped formation of life on planet earth.

In an earlier point we have discussed that there is a secret role of Saturn. The role is if there was no Saturn then nothing would have stopped Jupiter from colliding to earth and then

we would be just a part of the asteroid belt which currently lies between mars and Jupiter.

Similarly another important benefit of gravity is observed in the axis of our planet. The earth is tilted on its axis. This helps in making sure we don't have too much heat on any one side. If we didn't have this axial shift then earth would have scorched its surface just like Venus.

6. **Temperate Zones**

Some life has adapted to the most frigid places on Earth (including Antarctica, where the planet's record low was set at minus 128.6 degrees Fahrenheit or minus 89.2 degrees Celsius) and its hottest deserts (including El Azizia, Libya, where the record high of 136 degrees F, 57.8 degrees C, was recorded). But life achieves greatest diversity in more temperate climes, namely the tropics, where moderation rules

7. **The Deep Blue Sea:**

About 70 percent of our world is covered by oceans. The significance can't be overstated: Abundant liquid water is the most significant distinguishing factor on this planet that supports life.

This is not only because of the fact that biochemical reactions require water but also because without our oceans earth's temperature moderation would be an issue. Because of the high specific heat capacity of water, they can retain a very high calorific amount of heat.

8. Presence of lighting

The presence of charge in our atmosphere is one of the most important events in the chronology of time that happened to earth. Lighting helps in creating compounds of nitrogen like nitrogen dioxides and ammonia which are the initial stages in chemical evolution. These chemical later helped in creating methane and fundamental amino acids which are the building blocks of primordial life on planet earth.

The well-known Miller-Urey experiment in 1953 raised the possibility that lightning may have been a key to the origin of life.

9. The importance of green

We generally take the colour green around us for granted, but it is the most important factor contributing to the development of life. But without some very promising events chlorophyll as we know will never have existed in the first place.

In our earlier chapters we have discussed how elements are created inside stars. Our sun is not large enough to create all these elements so they are from other stellar systems. But the important element in chlorophyll is magnesium. If the element has not arrived at the right moment for the molecule to be created, photosynthesis as we know wouldn't exist. This shows how even a simple miscalculation in the entire equation of life could have made earth a barren wasteland.

10. Space

Earth doesn't exist in a vacuum. The space in our solar system is dotted with asteroids and comets, plus dust and traces of gas. Even now, small space rocks rain down on Earth daily. Big ones slam into the planet often enough to keep NASA and other space agency on constant lookout. And in the early years of the planet's formation, giant collisions with comets and asteroids brought water and other important chemicals to the planet, making the origin of life possible. If these collisions were not in proper order or in the later stages after the mantel got cooled, life as we know wouldn't exist.

A detailed discussed is already done in earlier chapters on this event.

11. Tectonic movements

One of the many unique features of earth is its tectonic movements. As we know that the time line of earth is divided into several periods like Hadean, Archean, Cambrian etc. In each of these time frames the face of earth changed many times before becoming as we know it today. These changes altogether helped creating many important features including hills, mountain ranges and seas as well as other environmental factors which together helped in supporting the evolving life on planet earth, from simple prokaryotic to eukaryotic life form.

12. The blessing of chance

While discussing several factors in a scientific community we never really care the importance of time and its blessing in the form of chance. The earth originated around 4.5 billion years ago and since then if chance to evolve was not present that life could not sustain. If a giant asteroid or meteorite crashed the planet during the initial stages, then the entire concept of life

as we know would be extinct from the face of earth.

So in conclusion to these several factors we can understand how important time is to bring life on any planet. This is the reason why the probability of finding life is so rare on any planet, because if any one of the factor is missing than life as we know wouldn't exist.

But while finding life we are always neglecting the most important factor which is time itself. Let us have a look how this factor contradict our findings. Let us say there is a planet on a distant galaxy about 5 billion light years away from earth. Let us consider that the species living on it are very advance and they used a telescope to look towards earth, what do you expect them to see?

The answer would be nothing as 5 billion years ago earth was not there. But this event is true for all celestial observation and this important drawback is the bottleneck to our technology to explore the space. The best alternative is to understand tachyon.

During both chemical and biological evolution there were significant events which led to certain specific chemical reactions which helped to steer the cycle of evolution in the right direction. In the next chapter we will have

a detail analysis of these biochemical changes and other significant factors and how they still affect us to this day and age. For the time being let discuss about tachyons.

Tachyon

To understand the mysteries of the universe completely, the only available option with maximum efficiency is travelling to the distant corner of our universe. But in order to become an interstellar species one thing which we have to overcome is the limit of time. Most celestial objects are millions of light years away from us and we won't have the necessary required time limit of that magnitude.

The solution to this problem was given by Albert Einstein in his theory of relativity. We have already discussed the bending of space and time in earlier chapters and based on that idea there are two ways by which we can venture across the universe. They are as follows:

> ➢ If we can bend space- time continuum with the help of gravity. This can be achieved if we can find massive dense gravitational bodies which will slow the time for us and we can venture to far distances using their gravitational

assistance. Under this concept comes the idea of worm hole also.

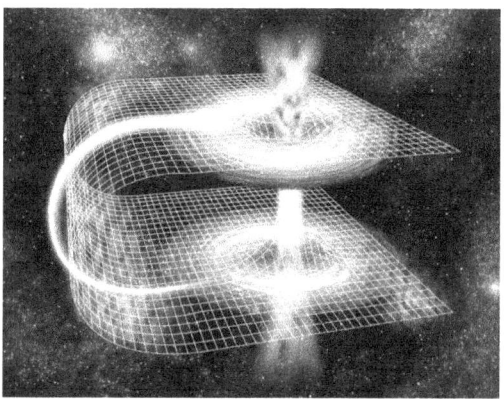

A worm-hole will bend space-time thereby creating a shorter route for a huge distance. Courtesy: space.com

Worm holes are theoretical and they are yet to be discovered. But with this theoretical concept several research works are being conducted across the globe to find a suitable solution to our problem of interstellar travel.

> ➤ The second alternative is travelling faster than the fastest thing known to human in our universe, light. By now it is very much clear that space and time are two sides of the same coin and the only thing which can transverse through it fast enough is light.

The concept on which theoretical physicist are working is to find a substance which is faster than light. The reason for this search lies in the fact that universe is expanding faster than the speed of light, which means that there are certain materials which are capable of this but yet to be discovered by us.

However most physicists believe that faster than light particles cannot exist as they will violate the fundamentals of universe, which state causality must proceed before result (Grandfather Paradox). In modern time we mostly use the term to refer to a mass field rather than a particle. The complimentary particles to tachyons are Luxon (always moving at the speed of light) and Bradyons (moving slower than the speed of light). Both of them exist in reality.

The rest mass of any object as per relativity while in a motion is given by the following equation:

$$E = \frac{mc^2}{\sqrt{1 - \frac{v^2}{c^2}}}$$

Hence an object moving at the speed greater than light will need infinite amount of energy.

Although the existence of tachyon particles in reality is a topic of debate in the realm of particle and theoretical physics, but that is the beauty of science. There was a time when scientists were arguing over the existence of nuclear fission and fusion and now it is taught at high school level.

No matter the result, these two ways are the only highlighted idea which might enables us to travel as interstellar species. However we are technologically far behind to implement or even detect these particles. Maybe not today but in some distance future, humanity may become an interstellar species.

There is a quote by Tim Fargo, "until you cross the bridge of your insecurities, you can't begin to explore the possibilities". This is applicable not only to our limitation of technology to find tachyons or creating FTL (faster than light) travel options but also in our life. So with that positive note let's continue to the next milestone of our journey.

Chapter 8

A symphony of beauty

Life is explained in a variety of ways in various cultures as well based the different school of thoughts. An artist may have a different concept of life in comparison to a biologist. Here we are discussing the general concept of life based on biochemistry and what does it mean to be alive.

Since millennium the concept of life and death has fascinated generations and every culture has some form of customs and rituals associated with them. But the big question lies what signifies life?

Scientifically speaking certain biochemical reactions which are undergoing inside an organism gives them the characteristic features of life, other than that the basic elements of nature like carbon, nitrogen and phosphorous are all same. Our discussion is not about individual chemical reactions supporting life because that will take several editions of encyclopaedias to complete. So let us discuss from evolutionary stand point the idea of time and how several incidents played like a fine symphony in their order of perfection to create the balance form of life as we know it.

In the earlier chapter we saw the various events which occurred in their precise order to create earth into a liveable planet but in this chapter we will focus on the topic from an evolutionary point. It is crucial to understand

that life as we know today originated in a rather simple format with time as its only sole protector. Without time, this life would never have developed into the complex ecosystems as we know today. So let us have a look into this network one step at a time.

Chemical Evolution

After the time period of cooling of earth's core and formation of a stable mantle, the offset of basic chemical reactions started to initiate in the early earth. These reactions are considered very simple in today's time but millions of years ago originating these reactions required the boom of time. Any celestial event would have disrupted the entire process. We will have a brief introduction of these reactions because it is physically not possible to analyse millions of reactions in details which occurred during the entire evolution process in one single volume.

The chemical evolution transcends through the concept of space and time, because without these fundamental life sustaining compounds, life as we know wouldn't exist in the first place. Several branches of science like astro-chemistry, biochemistry and prebiotic-chemistry are dedicated in the study of these compounds and how they have reshaped the world as we know today. Since the last few

decades, milestone development in scientific achievements has helped us to understand those prehistoric conditions better. This also helped us to solve the mysteries about genetic materials (DNA & RNA) with the help of human genome project.

We will understand chemical evolution using the following model approach.

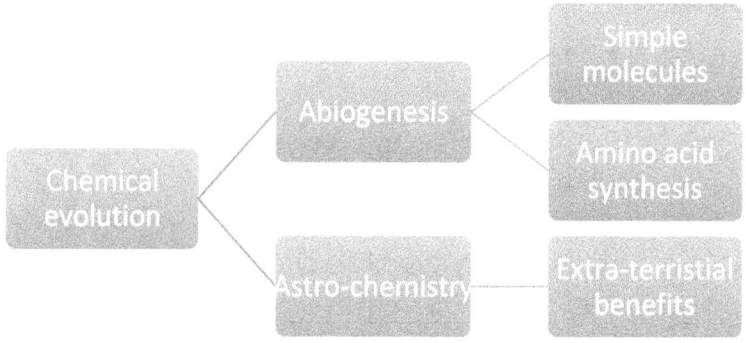

Miller–Urey experiment

This was the first experiment conducted to understand the chemical conditions which prevailed in the early earth, thereby understanding the origin of life. It was conducted at university of Chicago in 1952 by Stanley Miller and Harold Urey. This was one of the earliest confirmations of the theory of abiogenesis.

This experiment was conducted as a mean to prove early life evolution in prehistoric earth. The basic ingredients in the earliest earth were nitrogen, oxygen, hydrogen, carbon. These elements combine to give basic compounds like water, methane and ammonia. It is to be noted that water is not only created by combination of hydrogen and oxygen, some came from the Oort cloud as discussed in the earlier chapter. This experiment proved how the first amino acids were created which led to the development of DNA and thereby led to the biological evolution.

To conduct this experiment, water, ammonia, methane and hydrogen were kept in a five litre glass flask connected to a 500 ml water flask to induce evaporation like condition which was prevailing in the early earth due to its heated core condition. Lighting played a very important role in early earth as it helped in several important organic syntheses by providing the necessary activation energy. This is the reason why most ancient civilization and culture worshipped lightning. To replicate the same condition, high voltage of electric sparks were regularly released inside the reaction vessel.

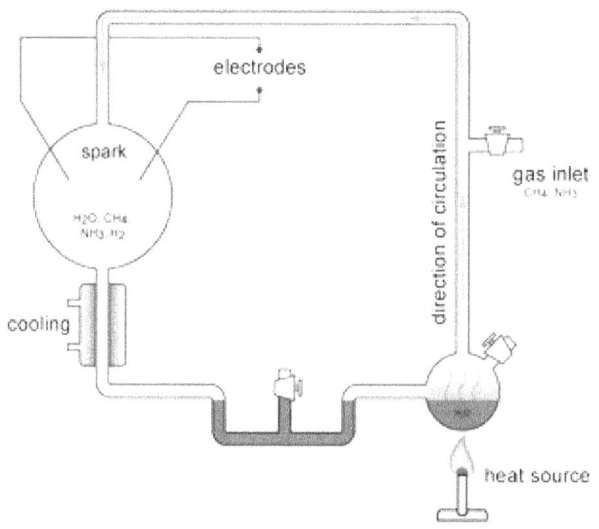

Initially the solution turns pink, followed by deep red as time pass by. To prevent microbial contamination mercuric chloride was added and to stop the reaction barium hydroxide and sulphuric acid was added. After the reaction is over, using chromatography technique 5 amino acids were identified by Miller, glycine, α-alanine,β-alanine aspartic acid and α-aminobutyric acid.

The chemistry of this experiment was quite simple but beyond the requirement of our discussion here. Later on several other experiments were conducted by several notable scientists like Jeffery Bada, K.A.Wilde etc. The idea showed with later modification that around 10 to 11 amino acids out of the known set of 20 can be proved to have originated in the early prehistoric world.

This experiment laid the foundation of evolution and the idea of how life as we know originated. But the interesting part of this story lies in the fact that we had time in our favour, one wrong meteorite hit or one comet misdirection and all the wonderful chemical reactions would have cease to exist.

But time did not stop its role at this time, the story continued further into the biological evolution phase, which we will discuss in the next part of this chapter. However we will point out some evidences and impact of time on biological evolution, because the general idea of biological evolution is well known to every *Homo sapiens*.

Biological Evolution

Evolution is not a new topic and we all have discussed quite extensively about it in our high school, which is why this part of the chapter is really about looking at the evidences of evolution and understanding how time allowed us to evolve from simple single celled creature to multi-cellular living organisms.

1. Biochemistry and Embryology

The very basics of our biochemistry are similar with almost all creatures. We can find similarities between tigers and humans to some extent and so as we simple arthropods. Very simple evidence in this regard is the ultrasonography of a growing foetus of some creatures.

Some scientists have a belief that maybe evolution led to a single creature formation which later modified based on its environmental condition.

Common ancestor- some scientist believe that many organisms may have a common ancestor because of how the embryos resemble each other

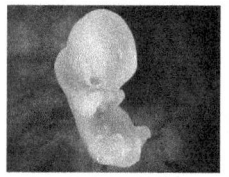 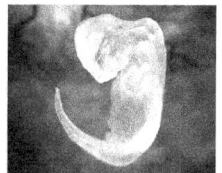 

- Puppy   dolphin   elephant

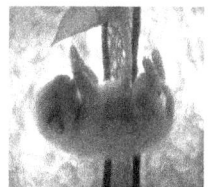 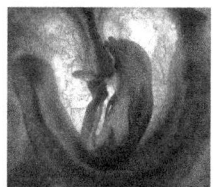 

 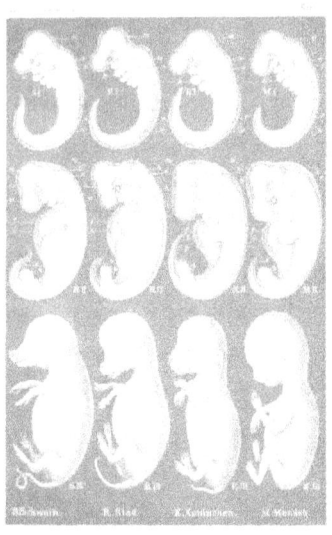

Fig: The data collected by Ernest Haeckel showing the comparison with different species including that of a human embryo during the gestation period

2. Biogeography

The geography of the planet changed a lot since its inception. This is clearly depicted in various scenarios like the presence of only 4 major blood groups across the world shows that once the entire landmass was united as Pangea. Groups that evolved since the breakup appear uniquely in regions of the planet, such as the unique flora and fauna of northern continents that formed from the supercontinent Laurasia and of the southern continents that formed from the supercontinent Gondwana. The presence of members of the plant family

Proteaceae in Australia, southern Africa, and South America is best due to their appearance prior to the southern supercontinent Gondwana breaking up.

Another example of this breakup can be seen in the presence of Marsupials in Australia only as a result of selective evolution.

But this puts a new perspective for us to discuss, evolution and life is not limited to our imagination. In our own planet there are millions of species living based on their environmental conditions, hence it is very much possible that the definition of life as we know is not applicable to the entire universe, we just have to expand our imagination. As once stated by Arthur C Clark, *"The only way to discover the limits of the possible is to go beyond them into the impossible."*

3. The Endosymbiosis Theory

This theory states that the organelles present in eukaryotic cells were once part of prokaryotic cells. The evidences for this theory were as follows:

1. Chloroplasts are the same size as prokaryotic cells, divide by binary fission, and, like bacteria, have Fts proteins at their division plane. The mitochondria are the same size as

prokaryotic cells, divide by binary fission, and the mitochondria of some protists have Fts homologs at their division plane.
2. Mitochondria and chloroplasts have their own DNA that is circular, not linear.
3. Mitochondria and chloroplasts have their own ribosomes that have 30S and 50S subunits, not 40S and 60S.
4. Several more primitive eukaryotic microbes, such as *Giardia* and *Trichomonas* have a nuclear membrane but no mitochondria.

A similar idea is the fact behind R.H.Whittaker's five kingdom classification; Monera, Protista, Fungi, Plantae and Animalia.

The only reason why it was possible for such a massive scale of evolution is that we had the luxury of time without which we would have never walked on our beautiful planet. But understand the unique idea of time and its relativity, for someone 100 million light years far from earth; dinosaurs are still roaming on earth and someone 700 million light year ahead will see a barren piece of rock.

Our biggest ally is time but at the same time it is also the biggest hurdle in our effort to explore the final frontier of space.

Chapter 9
Dark Matter - the mysterious anomaly

Dark matter and anti-matter are some of the few most fascinating words used by science fiction movies in almost every possible way. Although they are quite interesting in these movies, in reality they are much more mesmerizing and their existence puts our life in a perspective and our origin into question. If we are travelling back in time to our origin then it is crucial to discuss about them.

First of all understand one thing very important, dark matter and anti-matter are not the same. Matter as we know is what we all are, including our surroundings, our planet, and our universe. The subatomic particles like electrons, protons, neutrons are all constituent of matter. But this matter is a very small proportion of the total content of the universe.

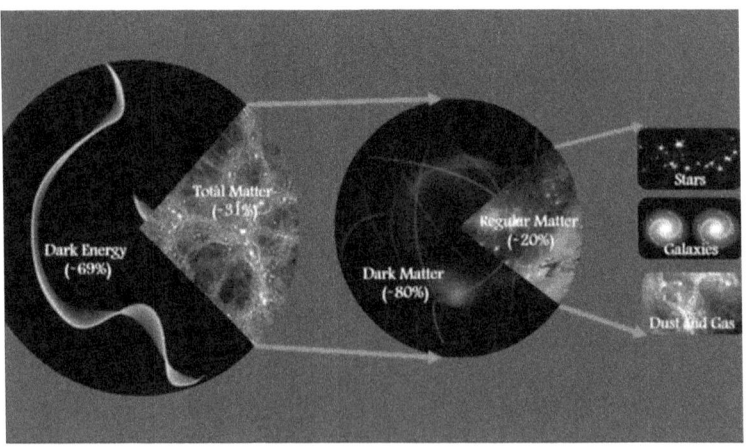

Antimatter has a few properties flipped, such as the electric charge. For example, the antimatter version of an electron is a positron. They both have the same mass, but have opposite electric charge. Antimatter is not as exotic as science fiction makes it out to be. For starters, antimatter has regular mass and accelerates in response to forces just like regular matter. Also, antimatter is gravitationally attracted to other forms of matter just like regular matter. Studies related to antimatter is extensively done and we now know that for every existing matter there is a counter antimatter. Scientists have isolated them and several studies based on boson particles and fermions are in continuation.

However this is not the case with dark matter and dark energy. In simple way, dark energy is the one which repel gravity. Scientists believe it to play a significant role in the expansion of the universe but proper understanding over this topic is still not available. It is crucial to understand that we are matter and so is light, hence all our approaches to interact with dark matter is futile as it doesn't interact with any matter. The concept of dark matter is not only limited to modern day science but is discussed in certain ways in many cultural and religious matter.

So if practical approaches are unable to detect dark matter, then how do scientists approach this topic? The answer lies in theoretical physics. For example, let us try with a very basic approach; a circular motion. Have a look at the diagram below, depicting a circular motion with two planetary bodies located at two different orbit around the sun. It is crucial to understand that according to physics the centripetal speed of the outer ring should be less than the inner ring.

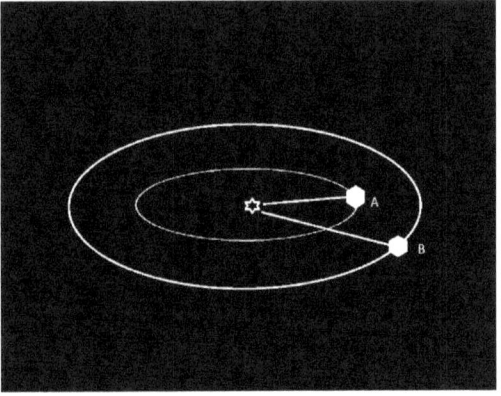

You can approach this via a simple experiment, while riding a Ferris wheel, the outer part of the ring moves slower than the inner part. However while comparing the rotation speed of stars on the exterior of spiral galaxies to that of the internal stars a shocking observation is made which shows that they have almost similar speed. The only logical explanation is that they are experiencing external unknown gravitation which is undetectable, most likely dark matter. It is very crucial to understand that dark matter will still possess gravitational field, the only difference is they are not directly detectable.

Dark matter could also explain certain optical illusions that astronomers see in the deep universe. For example, pictures of galaxies that include strange rings and arcs of light could be explained if the light from even more distant galaxies is being distorted and magnified by massive, invisible clouds of dark matter in the foreground-a phenomenon known as gravitational lensing. Currently leading groups of scientists, including the CERN research team is working to analyse these dark matters and unlocking the mysteries of the universe.

A fascinating revelation came in the 1990s when scientists discovered unique property related to dark matter. Earlier scientists used to think that gravitational pull will slow down the expansion of the universe. This however is scientifically inaccurate, because we have already discussed in the previous chapters how for any system it is vital for entropy to keep expanding. But expansion of universe is not the issue, the rate of expansion is. Based on scientific data the rate of expansion of universe is more than that of speed of light. The only plausible reason for this observation is the presence of external forces to accelerate the expansion. The more the universe expands the larger this force is becoming. Due to this mysterious nature, scientists called it dark energy.

This finding also proved that our universe is created after the death of a previous universe and this cycle will keep going on. This is because as entropy keeps on increasing the gravitational pull will become weaker and at one point the entire universe will collapse on itself creating the next big bang. This claim is further supported by Noble laureate Sir Roger Penrose, who using Einstein's theory of relativity observed spots in the universe with electromagnetic properties known as 'Hawking's Points". These are the remnants of a previous universe whose demise led to the creation of our own. It is part of the "conformal cyclic cosmology" theory of the universe, and it is suggested that these points are the final expulsion of energy called 'Hawking radiation', transferred by black holes from the older universe.

It is crucial to understand that the reason why we have to know more about this dark matter is because without knowing all the forces present in our universe, we cannot understand its mechanics and hence time travel will always be a dream. To understand dark matter we have to change our approach and scientists are working on several aspects of it. Dark energy is much more powerful and abundant than the form of energies we know in our universe and can be a fissile source of power to attain faster than light travel.

In the last two chapters of the book we will discuss two most promising and at the same time hypothetical scenario which defines our existence and makes us wonder the mysteries of our universe.

Since the very first chapter, we have started to explore the universe to understand its creation and to understand how time came into existence, but more important we understood the role time played in creating life on our beautiful planet. Time is a mystery and with the help of quantum physics, string theory and other branches of theoretical physics we have discovered quite a bit of information about it.

Although a lot more needed to be understood but that is the beauty of science as we know it. To keep exploring the unknown and to unravel the mysteries of the universe.

Chapter 10

Time travel - A journey to remember

Now after travelling through the origin of space and time and understanding the various scientific realities of cosmos we have reached the endgame. The theoretical information discussed in the previous chapter deals with the concept of time but these last two chapters are going to showcase the true power that time holds in its maximum potential. But before we proceed to actually dive into the beautiful yet daunting experience of time travel and multiverse, there are few important clearances which are required.

In both of these chapters everything that we will discuss will be strictly scientific in nature and we will break any myths associated with them. However it is to be always kept in mind that we are limited by our technology of modern era, maybe in future this bottleneck won't be there to stop us from harnessing the time travel. But I am sure most of you at this point of time must be thinking that if time travel ever becomes possible why is our future generation not coming back in time to help us? I can assure you by the end of this chapter you will have all your answers to every question you ever thought of regarding time travel.

It is very crucial to keep a flow of all the points in this chapter so as not to make it a mixture of confusion. To avoid this we will approach this

chapter in a very systematic way, first we will understand the concept of dimension and then we will clear any myths about time travel and the end we will discuss the paradoxes of time travel.

The dimensional approach

If you are a high school or graduate science student you already know what a dimension is but let us have a brief outlook on what do we define as dimension.

A dimension is a point we define in co-ordinate space to denote a particular axis. Now since we live in a three dimensional universe we mostly experience everything in 3D, which are length, height and width. Normally if I ask you to imagine any dimension beyond the third, it is going to be very tough on you. This is because our minds are not designed to think beyond third dimension; lucky for us we have quantum mechanics in our disposal.

One of the pioneer scientists of our era, Dr. Stephen Hawking did stunning research in the field commonly known as string theory in general and M-Theory in the scientific community. This theory in itself would require series of books to explain, however we are just interested in few of its finding for our discussion. As per the theory, we have already calculated almost ten dimensions mathematically, they are discussed under super-string and loop quantum theory.

But why are we interested in these dimension logic, to answer this question, I want all of you to look at your shadow.

I hope you all are looking at your shadow, what do you observe? Is it same as you are? Does it absolutely define your picture accurately?

Well, definitely not and there lies the answer to the proper understanding of dimension. This is because you are a three dimension being and your shadow is in two dimension. The same is applicable for your photograph as well. This technology is used by many face detection software which can differentiate between an actual face and a photograph.

I hope you understand why dimension is important but how does it affects time? We already know by now that time is the fourth dimension but there is more to the story yet. Now go back to the shadow example where it was clear to you that your shadow is not an accurate description of your looks, but now imagine you are a two dimensional being. You have no idea of what three dimension looks like and now you are observing the same shadow. For you it seems normal because you have no idea that a third dimension which is height even exist. In a similar fashion if an

extra-terrestrial life form of higher dimension ever comes to our earth or may have come to our earth already we would be completely incapacitated to render its true form. This is exactly what the true definition of god is in many religious texts. Just for example let us consider the holy bible. In the bible there is a mention of an angel type known as Cherubim or Cherub who are tasked to protect the Garden of Eden. The fascinating observation is that they are mentioned as animal-human hybrid creature, similarly in Norse and Hindu mythology. Why is there such similarity in their observation, or maybe they were observing something similar. I coined a word to describe such kind of life forms which are varied based on the dimension they belong to; it is called ***dimensional life paradox***. This paradox defines the concept described in the previous paragraph which proves why any creatures which are separated based on dimension can never interact with each other. Even if they somehow came in contact with each other, the creature from a lower dimension can never depict the true form of the higher dimensional creature. This is the reason why most religion says that we cannot physically comprehend the true form of God. If we consider any advance species as God than its true because they may be from a higher dimension which mathematics have proved using M-theory of existing the possibility of 10 dimensions.

Any sufficiently scientifically advanced being would be considered god-like by a lower class of civilization. I am not making any debate over the belief system but consider in a logical way, let us imagine if you can travel back in time just 500 years with all the modern era technology what will the people of that time consider you? I will leave this question for you to answer.

But the problem with a higher dimension is that in order to explore it we need to live in comparatively higher dimension. Meaning, if we want to explore time in the fourth dimension we need the fifth or higher dimension. Now you must be imagining that if that is the case, how can we navigate in three dimensions without any higher dimension? The answer is we are living within time; that is the beauty of this illusion which time creates. Let me explain with an example.

Let us say you asked your friend to meet you at your nearby park, so to properly give him the location you will need the co-ordinates of the park which includes the x, y and z co-ordinate points, where x and y are latitude and longitude points and z represent the height from the sea level. But do you think your friend will able to find you? The answer is no, because you never mentioned him the time to meet, so obviously he is not going to wait for you the entire day. So you see, we are using time as a dimension

in our daily life but we never realise its importance. Before we proceed I would like to let you know that geometrically speaking it is impossible to accurately demonstrate a four dimensional surface on a two dimensional paper, but still a pictorial representation is as follows:

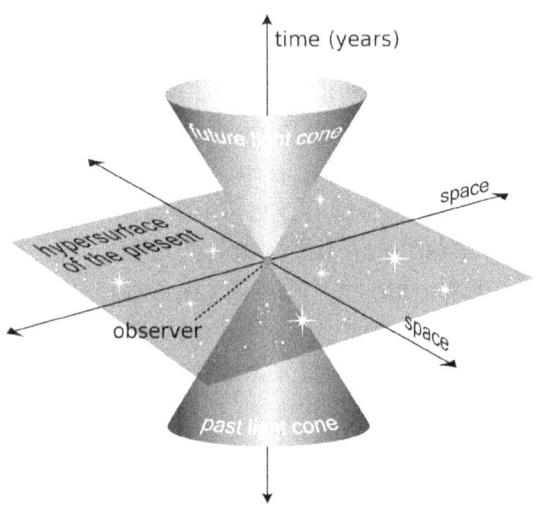

A light cone depicting the dimension of time. Courtesy: Forbes

So with the basic idea of what a dimension is and how time is a dimension we are at a crucial point to address some common myths regarding time travel and to understand the science behind it.

Myth 1: We can travel through time to go back to our past?

Let's start from simple logical reasoning; if I ask you to go to London can you go? The answer is yes because that place exists. But when we talk about travelling back in time in the past we have an infinite scope of possibilities. It can be 1 second, or 1 minute or 1 month or even a decade in the past. So in order to travel to these points in time these places must exist in the first place. Physically it is not possible to exist because then the randomness of the universe by virtue of its entropy will increase so much that the entire universe must collapse. Such an event of time travel is defined under the Fermi Paradox. However there is another alternative which we will discuss in the next chapter.

Another critical constrains for such a place to exist is that, even if a place like that exist which is the past of our present version, why will the same variables exists. Variables in this case are the human beings involved in any incident, a series of event following the same chronological order etc. For example let us say two years from now you and four of your friends went to a fair and brought strawberry flavoured ice cream. Even if there is another planet earth which is running two years behind us, the probability of having another you and exactly same four friends making the exactly same decision is scientifically an infinitesimal to zero.

REALITY

Although we cannot go back to our past, there is one thing that can be done to experience the past once more. In fact we experience this every day and have actually discussed in previous chapter.

If someday in future we can develop faster than light travel and can move 1000 light year ahead from earth within few moment of time, we can technically look back towards earth and observe the past 1000 years back. But that is all, just observation with a few modifications which we will discuss in the next chapter.

Myth 2: During time travel we cannot see our own past or future self

This misconception aroused in the first place due to several movie depictions which is not at all true. Any person is simply a variable in the grand cosmos and doesn't matter what you do to a variable unless it is a constant the equation will keep modifying itself. This is a basic mathematical approach. Hence this misconception is completely false. Maybe in distant future we can actually travel in future to meet ourselves or our next generation.

Myth 3: Time travel will be very common

This is the most vague point ever created, even if time travel or faster than light speeds

are theoretically developed their practical application will take much more time, and even in that situation it won't be accessible to everyone. This is because human body may never allow based on our biology to access that high level of velocity. Most of us can't even bear the hypersonic speeds, let alone speed closer to that of light.

PARADOXES IN TIME TRAVEL

Paradoxes are some complex situations which are purely theoretical but possess a significant logical response in analysing complexity of time travel. We will discuss few of the most significant ones here, although there are many more in discussion.

1. Ontological Paradox

It is also known as the casual loop paradox and is depicted in many science fiction movies but what does it depict?

This paradox suggest that a set of events will always repeat itself to give a same result every time. It is based on the rigid approach of time as per Newtonian frame. For example let us consider you gave an exam and you got 50% in the paper but now you want to go back in time to reappear for the exam and get more marks and lucky enough humans have developed sufficient to possess a time machine. As per this paradox this will be scientifically impossible, which is because time likes to keep itself unaltered. So when you go back in time maybe a new set of events will take place and the same result will be repeated.

However this paradox reveals a very major flaw in our human mind set; the concept of free will. As humans the biggest strong point we consider ourselves lucky enough to have is the concept of free will. Countless civilizations have fought with each other based on their free will. This was described by the Newcomb's paradox. Even modern world society is doing the same thing. But what if free will is an illusion and the final major outcome will always be there without any importance to the random variables. We have to explore the bigger picture and in that bigger picture where do we stand. We are a very small living being on a very small planet revolving a quite small star. This is why Carl Sagan always use to say that astronomy is a very humbling subject, it teaches you humanity and morality.

2. <u>Grandfather paradox</u>

This paradox is very well known and is also known by other common name as consistency paradox or butterfly effect.

The idea behind this paradox is that any change of event in the past will cause a ripple effect to change the present, sometime with drastic conclusion. The reason this paradox exists is because certain laws are fundamental to the universe including the idea of cause and effect. An action has to be performed first before a reaction is obtained. This is the reason why the timeline is always rigid and no attempt to change will be allowed by the universe because it is against the very law that holds reality.

Beside these paradoxes there is one more which we have discussed earlier which is time dilation. One of the most fascinating facts about time dilation is that it can be experienced in real time on a very small scale by the astronauts aboard the international space station.

As we reach the end of this chapter we have understood quite a bit regarding time travel but the few bumps that were there for time travel can be overcome in the next chapter where not only can we find the answer to the problems of time travel but at the same time we can answer

the most fascinating question of all time, are we alone in the universe?

Chapter 11

Parallel Universe - A closer neighbour

In all of known science, this particular topic which we are going to discuss is the most controversial one. With maximum iteration based on numerous fictional shows and movies, many confusing articles have made this topic a pile of mess. So let us sort through this mess and learn the reality. Is it really plausible or is it just a fiction and if it does what does it states about our universe.

The term parallel universe and multiverse are not same, although they are related to each other they are not the same word. A parallel universe is another universe which is parallel to our one but on the same plane so we cannot detect them based on traditional means. All such parallel universe combines together to give the concept of multiverse, which is a combination of all such parallel universe. This clearly shows the presence of multiple realities within the same plane in multiple universes.

Parallel universe concept is not a new topic of discussion in science, earlier recorded historical facts shows the use of similar concept in mythologies including many religion belief. The presence of a multiple universe is clearly mentioned in many tales of Hinduism in several instances of religious texts. In Norse mythology the gods were from a different universe and can transport between worlds. A

similar approach was observed in Greek and Egyptian civilization.

Even in philosophy the concept is widely used under a new name, "Possible World". But if parallel universes exist then there must be some scientific facts which can ascertain this hypothesis. So let us understand the concept of multiverse and the evidences associated with it.

WORMHOLE THEORY

This is the most basic theory associated with multiverse and without understanding this concept it is very difficult to fully grasp the idea of multiverse.

A wormhole is a theoretical gateway that can connect long distances within few seconds. For example consider this page as space and two diagonally opposite ends as two points in the space, may be two galaxies. Now if you were to travel between these diagonally opposite ends you will take a huge amount of time because distance is more. A wormhole is the shortest displacement that can be made by simply bending the paper and connecting both the ends of the space-time continuum.

A pictorial representation of the same concept is shown below which depicts how the space is bend under the influence of the worm hole leading to a shorter passage thereby allowing any advance civilization to explore the depth of space in matter of seconds:

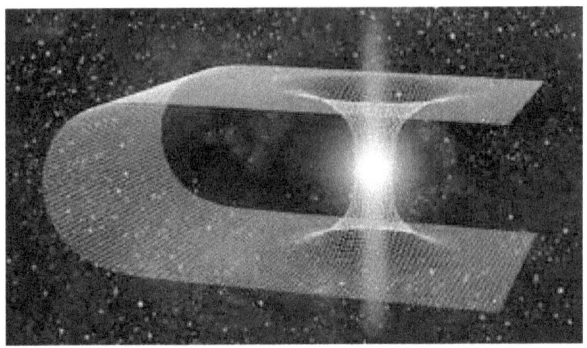

A worm hole

This theory was first postulated in 1916 based on Einstein's theory of general relativity. In 1935 Einstein and Nathan Rosen proposed a modified version of this theory that proposed that such worm holes can be used to travel between space-time as bridges. This theory later came to known as Einstein-Rosen bridge theory.

As per Norse mythology there was a bridge known as the rainbow bridge or the Bifrost that connects Asgard, the realm of Gods to Midgard, earth. A superhero movie series is also created based on this mythology which is one of my favourite of all time. Another such bridge in Norse mythology is the Gjallarbrú, which connect to the realm of the dead.

But by observing this shocking reality of similarities between mythology and scientific facts not only in this situation but in several others raise an important question, do our ancestors confuse a highly advanced civilization as Gods. I leave this to your own imagination to question.

The science behind multiverse

Multiverse concept can be approached via several routes including the relativity concept and EPR paradox or the idea of quantum mechanics or with the help of string theory.

Because this book is designed to provide a general understanding of the concept we won't drive too much in the scientific theories and mathematical derivations. This is because most of them require higher algebra, matrices and calculus along with an in depth knowledge of several physical theories. So let's keep the discussion interesting and simple so we can understand the science behind multiverse.

The quantum interpretation

We have discussed in brief previously about quantum mechanics. When dealing with any system, in this case it is the combination of several universes, quantum mechanics uses wave functions. They are mathematical tools which are used to define the various properties of the system. One of the most famous of such

wave function is the Schrödinger wave equation, which is widely used in physics and chemistry.

However we will discuss about the interpretation of multiverse based on quantum theory. To discuss this let us consider the famous Schrödinger cat experiment. If you put a cat in a closed cardboard box which is completely opaque and there are two vials present inside the box. One contains water, the other poison. The vials are linked with a machine which will only allow any one vial to be open after the pressing of the respective button. The cat can choose whichever button to press but from outside you don't know which button he pressed, until the box is open. So for you as an observant, until the box is open the cat is both dead and alive. This is called a state of uncertainty.

As per the many world hypotheses of quantum mechanics this is the situation with the different worlds in the parallel universe. Although they exist in parallel plains, there is no way they can interact with each other and will always remain apart.

Consider the idea of train compartments, each compartment is connected to the same train but they are physically separate from each other. Now if there are no windows in that compartment and there is no connecting

passage between them, it is impossible for the passenger of one compartment to know anything about the other compartment. This interpretation of multiverse is one of the simplest expressions among all the available ones.

The quantum interpretation is based on observer interpretation and directly question on what we deem as reality. In one world you may be the king of a country and in another world you may be an army captain. There are endless possibilities of computations occurring at an unprecedented scale based on the amount of data variations. There are many theories including decision theory, Born rule and many more governing the quantum interpretation.

String Theory and its interpretation of multiverse

String theory or M-theory is one of the most advanced theories in physics. The theory got its popularity recently due to its tremendous application in modern physics and also due to the world renowned physicist, Dr. Stephen Hawking.

As per string theory there are several evidences which actually prove the hypothesis of multiverse. In brief they are as follows:

1. **Level 1 parallel universe:** This is the simplest type of universe which can be possible under the multiverse theory. This type suggests that creation of universe follows the basic pattern of combination and permutation. As per Einstein's theory of gravitational waves the universe is finite but ever expanding. Using this logic there can only be a limited number of sets of arrangements possible using simple algebraic calculation. In that scenario computational model can help us to predict the similarity pattern of universe. In a research conducted by MIT, a crude estimate suggests that the closest identical copy of you is about $\sim 10^{10^{29}}$ m away. About $\sim 10^{10^{91}}$ m away, there should be a sphere of radius 100 light-years identical to the one centre here, so all perceptions that we have during the next century will be identical to those of our counterparts over there. About $\sim 10^{10^{115}}$ m away, there should be an entire Hubble volume identical to ours. This means there is an identical twin of earth somewhere in deep space with individual copies of each life of planet earth. Do you think on that planet, humans are responsible and have controlled pollution and global warming?

2. **Level 2 parallel universe:** By the 1970's, the Big Bang model had proved a highly successful explanation of most of the history of our universe. It had explained how a primordial fireball expanded and cooled, synthesized Helium and other light elements during the first few minutes, became transparent after 400,000 years releasing the cosmic microwave background radiation, and gradually got clumpier due to gravitational clustering, producing galaxies, stars and planets.

   Yet disturbing questions remained about what happened in the very beginning. Did something appear from nothing? Where are all the super heavy particles known as magnetic monopoles that particle physics predicts should be created early on? Why is space so big, so old and so flat (please do not confuse this with flat planetary surface hoax, this flat represent a basic Euclidian geometry), when generic initial conditions predict curvature to grow over time and the density to approach either zero or infinity after of order 10^{-42} seconds? What conspiracy caused the CMB (cosmic

microwave background) temperature to be nearly identical in regions of space that have never been in casual contact? What mechanism generated the 10^{-5} level seed fluctuations out of which all structure grew?

Inflation is a general phenomenon that occurs in a wide class of theories of elementary particles. In the popular model known as chaotic inflation, inflation ends in some regions of space allowing life as we know it, whereas quantum fluctuations cause other regions of space to inflate even faster. In essence, one inflating bubble sprouts other inflationary bubbles, which in turn produce others in a never-ending chain reaction. The bubbles where inflation has ended are the elements of the Level II multiverse. Each such bubble is infinite in size, yet there are finitely many bubbles since the chain reaction never ends. Surprisingly, it has been shown that inflation can produce an infinite Level I multiverse even in a bubble of finite spatial. Indeed, if this exponential growth of the number of bubbles has been going on forever, there will be an uncountable infinity of such parallel. In this case, there is also no beginning of time and no absolute Big Bang: there is, was and always will be an infinite number of inflating bubbles and post-inflationary regions like the one we inhabit, forming a fractal pattern.

There is also a debate of a third type of parallel universe based on superposition data. A general data of all types possible as predicted by string theory is presented in this diagram below.

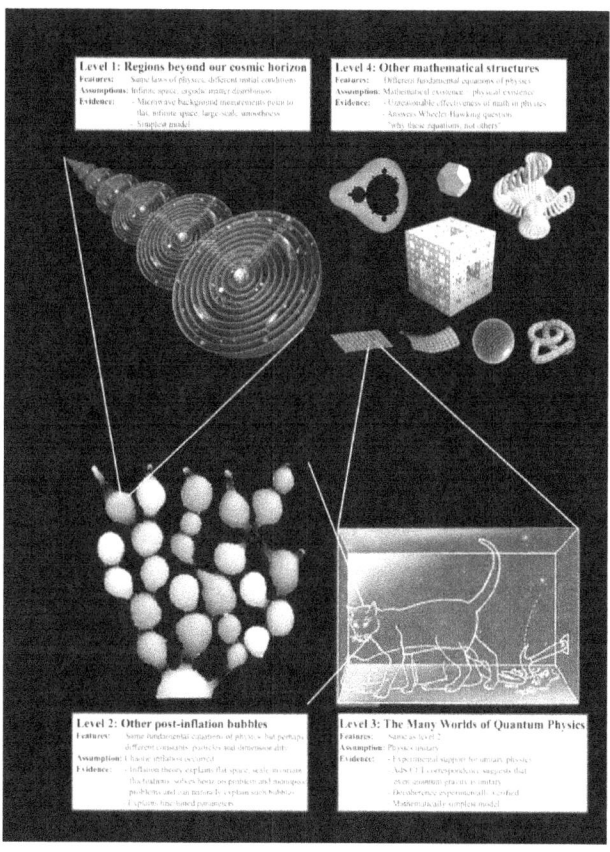

As per the modern form of multiverse theory the types of universe under this theory are as follows:

1. **Infinite universes**. We don't know what the shape of space-time is exactly. One prominent theory is that it is flat and goes on forever. This would present the possibility of many universes being out there. But with that topic in mind, it's possible that universes can start repeating themselves. That's because particles can only be put together in so many ways, restricted by their degree of freedoms.

2. **Bubble universes**. Another theory for multiple universes comes from "eternal inflation." Based on research from Tufts University cosmologist, Alexander Vilenkin, when looking at space-time as a whole, some areas of space stop inflating like the Big Bang inflated our own universe. Others, however, will keep getting larger. So if we picture our own universe as a bubble, it is sitting in a network of bubble universes of space. What's interesting about this theory is the other universes could have very different laws of physics than our own, since they are not linked.

3. **Daughter universes**. Or perhaps multiple universes can follow the theory of quantum mechanics (how subatomic particles behave), as part of the "daughter universe" theory. If you follow the laws of probability, it suggests that for every outcome that could come from one of your decisions, there would be a range of universes — each of which saw one outcome come to be. So in one universe, you took that job to China. In another, perhaps you were on your way and your plane landed somewhere different, and you decided to stay. And so on.

4. **Mathematical universes**. Another possible avenue is exploring mathematical universes, which, simply put, explain that the structure of mathematics may change depending in which universe you reside. "A mathematical structure is something that you can describe in a way that's completely independent of human baggage," said theory-proposer Max Tegmark of the Massachusetts Institute of Technology, as quoted in the 2012 article. "I really believe that there is this universe out there that can exist independently of me that would continue to exist even if there were no humans."

5. **Parallel universes**. And last but not least as the idea of parallel universes. Going back to the idea that space-time is flat, the number of possible particle configurations in multiple universes would be limited to $10^{10^{122}}$ distinct possibilities, to be exact. So, with an infinite number of cosmic patches, the particle arrangements within them must repeat — infinitely many times over. This means there are infinitely many "parallel universes": cosmic patches exactly the same as ours (containing someone exactly like you), as well as patches that differ by just one particle's position, patches that differ by two particles' positions, and so on down to patches that are totally different from ours.

Famously, physicist's Stephen Hawking's last paper before his death also dealt with the multiverse. The paper was published in May 2018, just a few months after Hawking's death. About the theory, he told Cambridge University in an interview published in The Washington Post, "We are not down to a single, unique universe, but our findings imply a significant reduction of the multiverse to a much smaller range of possible universes."

Everyday new evidences in particle physics and other branches of science is being discovered. During the final computation of this book, a new type of sub-atomic particles called muons. Many groundbreaking researches are being conducted in Large Hadron Collider and gravitational wave research being conducted at LIGO.

The beauty of science lies in its everlasting discoveries. So maybe new evidences regarding multiverse will arise very soon or maybe who knows a friendly alien will visit us and become a superhero for our planet. While ending this chapter, I would like to ask you a question. Now- a-days we are creating superhero stories based on science fiction data, what do you think ancient civilization did with their encounter with scientific abnormalities or phenomena?

Chapter 12

Our Future - Living the dream

When we started the journey from the first chapter we started the discussion on how universe got created and the creation of time and by the end of book we have learned quite a lot including time travel, parallel universe, supermassive black holes, Einstein- Rosen Bridge and many more. But do you really think these are modern time discoveries or just lost treasure?

Most ancient religions and civilizations of the world states about these mysterious observations in their respective genre. Maybe they were not able to define exactly what they were but they did observe them. The fault lies with us, instead of striving to learn about these mysteries we started to prove our strength and destroyed human race in the name of war and genocide.

Science is the language of universe, some may hate it or may even fear science. But you yourself is the best example of science then how can you hate it. Human beings are the best example of the most complex machines ever created by nature. The amount of biochemical reactions which are going at any given time is unbelievable. A small hormone of dopamine and serotonin can uplift your mood and attention and a decrease can cause depression. Everyday new evidences of a brilliant new article about science is getting

discovered so we cannot say that we are alone in the universe.

The basic ingredients of life lies in the periodic table of elements which are present across the universe, so if combination of these elements and grace of time can create life in our planet, life is there out there in our beautiful cosmos. Maybe there are some new elements waiting out there to be discovered.

Time and space is not just related to science but it defines how you consider your life, there are billions taking birth on our beautiful planet but only few are remembered because of their actions to advance mankind. The generation of renaissance, enlightenment age were the ones who ignited the journey of discoveries and it propagated till the information age. We have the maximum potential then our ancestors due to the limitless power of science and knowledge but at the end we are wasting this precious time. Just like any planet, each individual human has limited time, how they use that time totally depends on them. Most of the time, we are trying to prove ourselves to others and instead of growing, we are damaging our growth.

Science teaches you to be humble in its pure form. There are many people across the world who may have a background in science but has no relation with it. In my short span of life

as a student I have seen many who have joined as engineers but with no interest in the background, they just wanted a basic job with that degree. The problem with this attitude is that you will develop a hate for science because of this forced nature of the job. Science is not a subject, it is a thought process, and it makes you humble, secure and peaceful from heart. It removes the idea of grandiose narcissism and makes you realise the true meaning of humanity.

Humans explored the world, started from caves dweller to space dwellers and maybe in future we will become interstellar species but what define us is our humanity. The love we have for each other, the selfless sacrifice we make for mankind to reach new height. Our strength comes from our belief of hope, determination and perseverance. Maybe that is why universe choose to help us with the blessing of time so that we can decipher the true meaning of universe's message and become intergalactic species.

To a future, where mind is free, thoughts are pure and a world of harmony and growth. Let us hope we can sustain the journey to the final frontiers of technological advancement with the essence of humanity in our heart, so that our future generation residing in which ever planet can proudly say we are humans from earth.

I would conclude the journey with a beautiful poem by American poet Dylan M. Thomas which truly describe how we should live our life and utilise this time which we have been graciously given by the universe.

"Do not go gentle into that good night,
Old age should burn and rave at close of day;
Rage, rage against the dying of the light.

Though wise men at their end know dark is right,
because their words had forked no lightning they
do not go gentle into that good night.

Good men, the last wave by, crying how bright
their frail deeds might have danced in a green bay,
Rage, rage against the dying of the light.

Wild men who caught and sang the sun in flight,
and learn, too late, they grieved it on its way,
do not go gentle into that good night.

Grave men, near death, who see with blinding sight
Blind eyes could blaze like meteors and be gay,
Rage, rage against the dying of the light.

And you, my father, there on the sad height,
Curse, bless, me now with your fierce tears, I

pray.
Do not go gentle into that good night.
Rage, rage against the dying of the light."

References

1. Hawking, S. (2018). *Brief answer to the big question*. London, London: Hodder & Stoughton, Bantam Books. Chapter 2; General Theory of relativity
2. Hawking, S. W. (1993). Particle creation by black holes. *Hawking on the Big Bang and Black Holes,* 85-106. doi:10.1142/9789812384935_0005
3. Hawking's, S. (1988). *A Brief History of Time*. New York: Bantam Books.
4. Quigg, C., & /Fermilab. (2009). Lhc physics potential versus energy. doi:10.2172/963444
5. Shawhan, P. (2020). Available data products from Ligo-virgo gravitational WAVE Searches. doi:10.26226/morressier.5fb692d74d4e91fe5c54c276
6. Poggiani, R. (2021). Gravitational wave astronomy with compact binary mergers. *Proceedings of the Golden Age of Cataclysmic Variables and Related Objects V — PoS (GOLDEN2019)*. doi:10.22323/1.368.0052
7. Cern accelerating science. (n.d.). Retrieved April 13, 2021, from https://home.cern/
8. Loff, S. (2015, January 09). Solar system and beyond. Retrieved April 13, 2021, from https://www.nasa.gov/topics/solarsystem/index.html
9. Exoplanet exploration: Planets beyond our solar system. (2015, December 17).

Retrieved April 13, 2021, from https://exoplanets.nasa.gov/
10. Howell, E. (2015, January 30). How are planets formed? Retrieved April 13, 2021, from https://phys.org/news/2015-01-planets.html
11. Ollhoff, J. (2011). *Norse mythology*. Edina, MN: ABDO Pub.
12. Feinberg, G. (1967). Possibility of faster-than-light particles. *Physical Review, 159*(5), 1089-1105. doi:10.1103/physrev.159.1089
13. Chapter ten: Newcomb's Paradox. (1991). *Divine Foreknowledge and Human Freedom,* 205-221. doi:10.1163/9789004246683_012

- ***Special thanks to the online resources available under public domain by MIT and Stanford university***

www.ingramcontent.com/pod-product-compliance
Lightning Source LLC
Chambersburg PA
CBHW052353220526
45465CB00003BA/1090